T0206241

Effectively Managing the Case for Safety

This book examines how safety failings during the use of any designed product or system—be it a car, a building, or a chemical plant—can be mitigated through effective understanding of the conditions and controls surrounding its use.

Drawing on historical failures and their own real-world experience, Dr Andy Painting and David England explain how corporate culture, engineering safety, personnel selection, and proper safety auditing are key ingredients to maintaining safety in all aspects of an organization's operations. This effective strategy is also crucial to linking back to the design of future products in establishing where operational failures have been identified and can therefore be "designed out" in future iterations. The book challenges silo thinking among the various safety-related disciplines and shows how this can be counter-productive to effective safety management.

Effectively Managing the Case for Safety draws on key features from engineering, design, and health and safety processes, which, when used cohesively, promote a better working environment for everyone and help to reduce wasted time, money, and effort for any organization. Safety is tracked from the initial design stage through any product's entire service life and includes evidence of how safety affects, and is affected by, all those who interact with a product, system, or project. Following their first book, *An Effective Strategy for Safe Design in Engineering and Construction*, which demonstrated how current construction regulations can be used as a framework to ensure that safety is embedded into the design of virtually any product from machinery to buildings, this follow up book defines what safe is, how it is initially derived, and how the operational safety of any product, during its in-use phase, can be managed and assessed. The result is not only to ensure compliance with relevant regulations but also to actively ensure the ongoing safety of all those who interact with a product or project.

Dr Andy Painting began his working career in the Royal Navy serving 23 years as a submariner. As a weapons engineer, his job involved tracing and rectifying faults with various components of submarines, as well as helping to design and test new and modified weapon systems. It was during one repair task, whilst hanging upside down trying to reach a valve that needed replacement, he began to wonder if the design of critical components could be improved. Endeavouring to find an answer, he took degrees in engineering and health and safety, which led to achieving a doctorate in 2015. The thesis for this doctorate was catastrophic failure in complex systems, which led to the creation of an administrative early warning system which can predict the early failure of virtually any type of engineering installation or system.

After the Navy, he became the Chief Engineer for His Majesty's Naval Base at Portsmouth, where he was in charge of all the infrastructure supporting the Royal Navy's home dockyard. This included manoeuvring out of shipbuilding halls, a component part of the Queen Elizabeth aircraft carrier. This brought the 600-year-old tradition of shipbuilding at Portsmouth to a close. In 2015, he left the dockyard to set up his own company providing engineering and health and safety advice to a range of clients including government, local authority, and military organizations. This advice is inherently based on an understanding of how to implement engineering solutions that reflect best practice with cost efficiency, balanced always with the safety of the end user in mind. He is a Fellow of the Safety and Reliability Society and a Chartered Engineer.

David England joined the civil service after leaving school, quickly learning administrative and management techniques as well as discovering a passion for the law. In the late 1980s, he was involved in the supply and development of fibre optics which, at the time, was a nascent industry with very little regulation or standardization. This involved working with government bodies, scientists, and some of the largest component manufacturers in the world. For 25 years, he found himself in operational management in such industries as distribution, retail, and petrochemicals. In 1999 he was asked to complete a risk assessment, without any guidance, of a facility he managed containing some 100,000 litres of petrol, and he wondered if health and safety was all it was cracked up to be.

Determined to find out, he undertook the prestigious NEBOSH Diploma in occupational health and safety and, in 2013, started a career as a consultant. This led to contracting as a CDM Coordinator working at HM Naval Base Portsmouth. Working on large, complex, and often high-value projects gave him a unique insight into the demands of the client, designer, and contractor, all of which had to be balanced against ensuring that the safety of the operatives remained paramount. Working in some extremely high-risk environments also influenced his opinion on providing practical and pragmatic safety advice. He is a Fellow of the International Institute of Risk and Safety Management. Both authors are Fellows of the Institute of Construction Management.

Effectively Managing the Case for Safety

David England and Dr Andy Painting

Routledge
Taylor & Francis Group

LONDON AND NEW YORK

Cover image credit: © David England, Dr Andy Painting

First published 2023
by Routledge
4 Park Square, Milton Park, Abingdon, Oxon OX14 4RN

and by Routledge
605 Third Avenue, New York, NY 10158

Routledge is an imprint of the Taylor & Francis Group, an informa business

British Library Cataloguing-in-Publication Data
A catalogue record for this book is available from the British Library

ISBN: 978-1-032-28462-0 (hbk)
ISBN: 978-1-032-27128-6 (pbk)
ISBN: 978-1-003-29692-8 (ebk)

DOI: 10.1201/9781003296928

Typeset in Futura Std
by KnowledgeWorks Global Ltd.

Contents

Foreword

The past half-century has seen an amazing spread of "risk thinking" throughout the industrialized world and, with it, a profession and industry of risk management services, software tools, and products. This trend is well into its third generation of practitioners. However, many would agree that for some time, there has been a sense of lost direction and inefficiency. Too often, risk assessments are done slavishly to meet corporate or statutory requirements, and the safety documents generated are hardly ever read, still less do they live up to the aspiration of being "living documents".

Sadly, many corporate managers think they have protected themselves and their company if they have completed all the paperwork and delegated responsibilities to subordinates or contractors. That is far from the truth, and the reasons are simple: if an accident actually happens, all that paper may not only prove to have been inadequate but it may even self-incriminate, especially if hazards had been recognized but not controlled. It often looks deficient in the searching hindsight of an accident investigation. So, the name of the game is: do not actually have the accident. And that is what is meant by the word "effective" in the title of this book.

Of course, not having any accident, or being totally "safe", is an ideal that cannot be reached, but the same focus and commitment should be applied to implementing and maintaining safe workplaces and equipment and safe operating procedures, as has been applied to the identification of hazards and assessing the risks. The follow-through should be clear and direct and communicated in a way that seems practical, realistic, and important to the people who have to do the work.

As a former risk assessor turned accident investigator, I have seen how risks actually crystallize into accidents and the devastation that these events cause to individuals, families, organizations, and careers. Much of my work has been in the context of the voluminous litigation that often follows. That is predominantly concerned with apportioning liability, which requires detailed analysis of the chain of causation. It is very often found that a hazard that has actually occurred had been recognized previously, but the risk had been judged "tolerable", or the safeguards that had been identified had not been funded, or the reasons for the safeguards had been forgotten, and so they fell into disuse.

To protect ourselves and our organizations against such pitfalls calls for a form of eternal vigilance. That needs to be thoughtful, carried out competently, and with conviction, but it should not be burdensome. It needs to be a constructive part of a management process of continuous improvement and a welcome and well-understood feature of daily working life. It is this crucial phase in the management of industrial risks that is the subject of this book.

Dr Tony Cox
MBE FIMechE

Preface

After working together at the dockyard in Portsmouth, Dr Painting and David England realized that their respective professions—engineering and health and safety—were looking at the same problems from opposite ends. Engineering tries to find solutions to making things safe when people use them, whereas health and safety tries to find solutions to the risks that exist in using any type of equipment. They realized that if only engineering and health and safety professionals spoke to each other, they could find better, more cost-effective solutions which would not have to rely on lengthy risk assessments in the workplace. This would involve designers being more aware of the operational environment in which their creations would be used, and safety professionals bringing their knowledge early on to a design project. It is this knowledge and expertise that now allows them to offer a principal designer service to many infrastructure projects under the company name Attis.

They developed a concept around the phrase "safe to operate, and operated safely" and set about writing the book *An Effective Strategy for Safe Design in Engineering and Construction* (Wiley) in order to explain this concept to designers, project managers, clients, and safety professionals.

In 2019, Dr Painting and David England were invited on to the design team for the Get It Right Initiative (GIRI). In 2020, they were both invited to sit on the competency working group of the Institute of Construction Management in order to investigate and develop ways to improve the identification and transferability of skills and competency in construction. In 2021, Attis was the first ever organization to be awarded the highest level of competency on the ICM's register of principal designers. In 2022, they were invited to present a position paper to the Industry Steering Safety Group with reference to the Building Safety Bill. This Bill proposes to improve competency in construction following the Grenfell tragedy. It is this experience, combined with their respective knowledge of engineering and operational safety, that led them to develop a series of books which demonstrate how effective safety, in any type of industry, can be readily established.

This book looks at how effective health and safety can be implemented and what changes could be made to improve safety philosophy.

A third book examining the future of health and safety has been extensively researched over the last year by the authors, and it is hoped this will be published soon.

The authors wish to thank Dr Andy Slade, for his kind advice and support in bringing these books into fruition in the first place, and Dr Tony Cox, for kindly agreeing to write the foreword.

1 Introduction

National laws are often created as a result of some situation or occurrence that is deemed, or has become, societally unacceptable. In the UK, the situation is slightly different in that there has been a predominance of common law for almost 1000 years. Much of this common law has, since the assertion of Parliament during the Industrial Revolution as a single source of new law, become formalized as statute law. Modern laws regarding people's health, safety, and welfare are exclusively statute laws, with an early example being that which concerned the health and morals of apprentices being passed in 1802. Common law derived from judgements made by *justiciae errantes* (wandering justices) who travelled all over Britain attending what were known as "circuit courts" in an attempt to centralize legal decision-making and move it away from local customs. These circuit courts were often engaged in cases where one citizen was bringing a complaint against another. It would have been unlikely, however, that anyone would have sought justice for a safety-related case due to the inalienable division between master and servant, landowner and worker. And even in later times, the reason that laws regarding welfare were made statutory is perhaps because there will always be those who are prepared to risk the safety of others in the name of their own profit.

With one great exception then, safety-related law is almost always reactionary: post-event; reactive. The exception, the Health and Safety at Work etc. Act 1974 (HASAWA, 1974), was so exemplary and far-sighted that, nearly 50 years later, it has only seen one section revoked (that being connected to the environment) and is the principal law under which most workplace safety-related offences are tried. But laws are only part of the story. Society today is, thankfully, far more concerned with the health and welfare of individuals and abiding by the law only prevents one from being prosecuted. Of greater import today is the ideal of creating safe working environments *because that is the right and proper thing to do*, not just because some dusty old parchment says so. This is what was envisaged by the 1974 Act: employers taking the time and effort to understand the risks in their undertaking and providing the necessary arrangements to prevent harm from occurring. Clearly, things do still go wrong, but as any accident investigator worth their weight will tell you, invariably this is because one of those provisions was not in place: either the undertaking was not fully understood or the necessary arrangements were not in place. Of course, in some circumstances, there is simply a plain disregard for the law, or worse still, a disregard for a proper duty of care.

DOI: 10.1201/9781003296928-1

In 2014 for his PhD thesis, Dr Andy Painting chose to study ten catastrophic events in such industries as oil, gas, nuclear, rail, air, and space exploration. His intention was to find common linkages in the causality of the events and, by doing so, examine if it was at all possible to *predict* failures that lead to substantial harmful outcomes. He noted that despite various tools and systems that monitor complex engineering environments, harmful events happen regularly in all types of industries. The solution he proposed in his thesis was an intelligent system that could monitor certain indicators, regardless of the complexity of the system and the industry sector. The system was designed to predict the potential situations that can lead to catastrophic events, thereby reducing loss of life, financial loss, and damage to property.

The result was a near-real-time monitoring and trend analysis of a number of operational and administrative factors, which Dr Painting referred to as "traits". By examining and analysing these traits, he discovered it was possible to produce an intelligent system that could predict catastrophe. One of the surprising outputs of this research was the need to standardize investigative reports and audits of existing systems. The type and style of communication used, combined with looking for specific condition levels for each of the traits would provide a truer picture of asset condition. This could then allow for informed funding prioritization for such things as maintenance, repair, and replacement. That is to say, the system would show which elements of an organization were closest to a predicted point of failure, providing the organization with a pragmatic indication of where effort (in terms of maintenance, repair, or investment) would be best focussed. It could also be used to helpfully disseminate the system's findings to all relevant stakeholders.

An abstract for Dr Painting's thesis suggested that an intelligent monitoring system could:

> monitor and predict potential situations that may, in the future, lead to a catastrophe. Based on documented lessons learned from past significant catastrophic incidents, the combinations of failures that have led to disasters can be grouped under three distinct engineering governance headings of "people," "process" and "tools." The intelligence uses data mining software to detect patterns in the collected information. The predictive capability is enhanced using trend analysis—and through the adaptation of this information using pre-processors—to find the type and amount of variance necessary to create a recognizable warning pattern. The type of information to collect and the particular data mining tools to use will be the subject of further research and testing.

Later, whilst working with David England as principal designers on some large, complex public infrastructure projects, the pair realized that practical and pragmatic safety needed to be developed not just across all industries but in the functionality of organizations within those industries. Returning to Dr Painting's thesis, they began to develop the monitoring system to encompass the wider safety aspects that any organization could face. The "people", "processes", and "tools" originally determined as headings were increased, and the subsections in each one were developed in line with both established and emerging safety-related thinking. The first aspect they wanted to tackle was the implementation of safety as an embedded facet of the design process due to the realization that safe operation of any physical thing—a product or a machine, for example, or a less-tangible thing such as a system, process, or organization—cannot be properly and safely operated if safety has not been intrinsically *designed in* to that thing in the first place. The outcome of those observations is the subject of their book *An Effective Strategy for Safe Design in Engineering and Construction* (England & Painting, 2022).

During the research for this book, and in general observations of the systems and organizations that the authors have worked both with and for over many years, it was noticed that although the concept of risk assessment is ostensibly well-understood, there was less acknowledgment or understanding of the issues that surround those assessments. Dr Painting's thesis determined some of the background factors that had direct or indirect causal effects on catastrophic industrial events. These factors, along with many others the authors identified, also have an effect on the control measures that are introduced by general workplace risk assessments. Understanding these effects—and identifying the causal factors that create them—is central to this book's intention.

In the previous book it was discussed how getting the right information to the right people at the right time was a fundamental part of delivering a successful project. We also pointed out that the *requirement* for safety, at every level, is intrinsic on a human and ethical level, as well as having been a legal requirement in the UK for decades. Ensuring something is made safe, and remains safe during operation, is not necessarily a call for expensive or outlandish solutions. Simplicity and pragmatism are invariably the most suitable solutions when it comes to the realistic application of safety.

In this book we shall examine what we mean by "safe" and how we can achieve it in whatever undertaking we find ourselves. We shall look at the myriad possible inputs to an undertaking that require consideration and assessment and, most importantly, how they can be assimilated, administrated, and audited. We will also look at how the control measures that we put in place to reduce risks in organizations can themselves be flawed, thus reducing the potential effectiveness of those controls. Understanding how and where these flaws might arise is crucial in developing more robust systems of risk control and management, and this stems from analysis and research that the authors have undertaken both independently and jointly. This has led us to create a simple mnemonic that breaks down these potential flaws into four basic categories, which anyone charged with the provision or management of safety can readily adapt to their own situation.

We pride ourselves that modern industry, and indeed much of our modern world, relies on information, to the extent that new ways of processing and disseminating information have been dubbed new "industrial revolutions". But information has always been the key to apposite decision-making. Surely King Henry V needed as much information as possible about the size and distribution of the French army and the lie of the terrain that faced him at Agincourt in 1415. But no one described the eventual move from messengers on horseback to using the telegraph as a "revolution". Instead, the way information flows between sender and recipient has been one of *evolution* rather than *revolution*: messengers, postal service, telegrams, telex, fax, email—they have all incrementally improved the method of sending information without changing the need for information in itself.

We shall also look at the maintenance of safety, or safe operation, of whatever it was that we designed and procured under the guidance of the first book. And the reasons for this we have already covered: sometimes things fail because of poor design; other times they fail because of poor management. And this is a possible outcome in any size of operation, or any type, size, or style of device or equipment. It is also extremely relevant at this time in the UK as we witness ever more regulation, especially in the construction industry, regarding safety and safe design, essentially in response to various harmful and tragic events of late. But perhaps what some of this new regulatory burden has missed is the fundamental causality of those events and, in particular, the flaws in previous control measures that were in place.

Size, as has often been said, does not matter. Whatever type of industry you hail from, and regardless of what sort of undertaking your organization is involved in, this book will help you understand what sort of information will help to keep your organization—and, more importantly, the individuals in it—safe. Whether you are the CEO of a multinational conglomerate or a business owner with a multitude of roles to fulfil, or a manager of a building, farm, manufacturing company, or airfield—this book will demonstrate how (and where) to look for the right information and, most importantly, how to use that information effectively.

2 What is "safe"

2.1 What does it mean

We may feel we understand the meaning of "safe" and what the concept of safety means to us and our work. We apply safety to what we design, manufacture, and use, whether consciously or otherwise. But does that, by definition, mean that which we do not consider to be safe is, therefore, "unsafe". But is there value in that assumption? The Oxford English Dictionary states one definition of safe as "not likely to cause or lead to harm or injury; not involving danger or risk". The key phrase in this definition is "not likely to" and is key because it requires further definition. Not in dictionary terms, but in terms of the person appraising the "safeness" of the thing or situation. But why should we need further clarification of how or why something is safe when there are situations when clearly something is not safe? For clarity, we should also cross reference "unsafe" and "dangerous". It may be argued that unsafe has a slightly more definitive meaning to not safe: unsafe suggests that harm is *likely* to occur, whereas not safe suggests, perhaps, that harm *might* occur. Dangerous, on the other hand, indicates that harm *will* occur, it's just a question of time. In terms of safety, we should avoid the use of dangerous as an adjective because it suggests that we have allowed such a situation to develop. In safety engineering, however, some situations call for this description—for example, in the case of runaway chemical events where the mixing of two seemingly benign chemicals can lead to a rapidly evolving energetic reaction. The speed of such reactions determines whether they are classed as detonations or explosions; either way, it would allow little time to decide which adjective would have been the most appropriate to describe them.

Placing one's hand in the flames of a bonfire is unsafe. It will, almost without doubt, lead to harm. Falling from height is unsafe as the effect of gravity is insurmountable, as is the inertia built up by a human body in freefall prior to reaching the ground. But what of crossing the road to catch a bus? Or travelling by train or plane? Travelling by public transport may not invoke a sense of danger that the initial two examples would, but injuries and even fatalities can occur here too. The important distinction is the *rate* at which injury or harm could occur. Although there have been instances of people who, having fallen or jumped from an aircraft (either without a parachute or whose parachute had failed to deploy) have survived the fall to earth these are very

DOI: 10.1201/9781003296928-2

much in the extreme minority of cases. We can say that falling from a plane is *almost certain* to result in fatality. Similarly, of the many millions of journeys taken by train in the UK, very few result in a fatality, and therefore we might say that travelling by train is *almost certain* to result in no harm occurring.

An individual's perception of safe can be one of expectation rather than confirmation. In some situations, the very real presence of a harmful consequence is considered a risk worth taking in order to realize a better—or more "safe"—alternative. This can be seen in the dangerous journeys taken by people fleeing war-torn or oppressive countries and regimes in order to establish better lives in another country. The migration of people is something that has occurred since the earliest human history, through two World Wars to the modern era. The strength and depth of the desire to flee conditions which are felt to be too dreadful to endure can probably only be truly understood by those who have regrettably experienced them. Any dangers clearly evident in the prospect of undertaking such a journey are set aside in the mind of the person doing so because, in relative terms, they are outweighed by the clear and present dangers of remaining where they are.

In less emotive ways, we can also consider those individuals who place themselves repeatedly in the presence of danger: soldiers, firefighters, and lifeboat crew are but a few examples of such individuals. In these cases, the term "safe" should probably be replaced with "least dangerous". These types of individuals, and their responses to danger, were considered by Hale and Glendon (Hale & Glendon, 1987) in their model of behaviour which addressed the factors of motivation and intention. This is perhaps also countered by the individual's level of confidence, not only in themselves but also in their kit and their comrades. Confidence in oneself can, however, lead to over-confidence or complacency, which are attributes that have no place in team-work, as one would find in the fighting of a war or a fire or rescuing someone at sea. Again, the Hale and Glendon model considers the *system* surrounding such behaviour and includes such things as the environment, the levels of training, the support structure, and others immediately in support of the individual. This is why soldiers, firefighters, and lifeboat crew all train relentlessly in real-world situations to hone not only their own skills and experiences but also their comradeship and group support. The knowledge that "someone has your back" can be a powerful motivator in being able to face extreme situations with confidence.

These are extremes, of course, but what about something less so? The practice of changing a lightbulb by standing on a chair is probably one that many of us have attempted, and by and large probably without too much cause for concern. However, we understand that chairs are not designed to be stood on, and we know from experience—usually from falling out of a tree as a child—that hitting the ground after a fall is in all probability going to hurt. We are also aware that, at our place of work, performing such a task would be forbidden—or, at least, it absolutely should be. We are also possibly aware that the majority of accidents occur in the home. And yet we perform this potentially dangerous task perhaps because on the rare occasions we have done so in the past, it has ended without consequence. We are *almost certain* that nothing untoward will happen.

But what do we deduce from comparing the seemingly benign situations of travelling on trains and changing lightbulbs to the desperation of fleeing a war zone? We see that safety is relevant to an individual's *perception*—their perceived understanding of how safe something is and the conditions surrounding it: the situation and its environment, if you will. This perception is nurtured from a variety of inputs and is further affected by a number of conditions relevant at the time upon which that interpretation of "safe" has to be made. Primary in the case of nurturing this perception is the disposition of the individual themselves. Some individuals are predisposed to have great belief in their own abilities and character. They tend to be assertive, outgoing, dominant characters who do not recoil from potentially dangerous situations, not because they do not understand danger as a concept, or even ignore it, but rather that they have an innate belief in their own capabilities in overcoming it. Explorers such as Sir Ernest Shackleton and Sir Ranulph Fiennes are typical examples of characters who undertook dangerous adventures that many lessor mortals would almost certainly baulk at. Similarly, people who become, for example, soldiers (especially those who go on to fight as special forces soldiers) and firefighters are generally imbued with this strength of character. But these two examples are also demonstrative of another set of factors that affect the perception of safety.

Explorers are, by dint of their title, rarely trained for their role. Certainly, they can be well equipped and well prepared, but the actual training is a little more "on-the-job" than perhaps most of us might prefer. With reference to Shackleton, whose exploits were part of an era known as the Heroic Age of Antarctic Exploration, he was very much at the cutting edge of what was known and what was possible at the time. Pushing the boundaries of current knowledge and invention and of facing unknown eventualities in hostile conditions may be the mark of many heroic explorers throughout history but is hardly the expectation of the ordinary working person. In the case of soldiers and firefighters, it is possible to assume that they, too, will not be prepared for *every* eventuality that they are required to face, but their respective training regimes will at least provide them with the tools to make informed decisions. Allied to that training is experience—the more battles a soldier fights and the more fires a firefighter attends, the more relevant experience he or she accumulates in assessing the safety or otherwise of the situation confronting them. Training, then, is not something we can simply adopt in isolation to provide safety. It is inextricably linked with skills, knowledge, aptitude, and experience, and we abbreviate these factors as SKATE.

Perception can also be affected by a type of cognitive bias called availability heuristic, where an individual bases their opinion or judgement on the ready availability of information about a particular subject. An example of this is the widespread negative press associated with the nuclear power industry when compared to, say, power generated by fossil fuels. Studies demonstrate that the number of fatalities per unit of power produced (that is to say, the rate) is drastically lower for nuclear power compared to fossil fuel; especially coal (Ritchie, 2020). Public perception, however, is clouded by media coverage of nuclear accidents and the storage of radioactive fuel. Even the estimated tens of thousands of premature deaths caused by, for example, the

Chernobyl accident in 1986 is many magnitudes smaller than the tens of millions of deaths caused through poor air quality as a result of burning fossil fuels. And this does not include the untold human suffering that will transpire from the inevitable climate change that now threatens us due to the profligate use of carbon-rich fuel sources. The German government decided in 2011 to end the use of nuclear power for commercial power generation and began a programme of phasing out the nation's nuclear power fleet (Brendebach, 2017). This was in direct response to the Fukushima Daichi incident in Japan in March of 2011, despite that event being caused by very specific circumstances unlikely to ever be encountered in Germany. In 2020, nuclear power contributed 12.5% of electricity generation in Germany compared to 24.1% by coal and lignite (Germany's Power Mix 2020—Data, Charts & Key Findings, 2021). In the UK, coal accounted for just 2% of electricity generation in the same year. Lignite, of which Germany has vast reserves, is much closer to the surface than "hard" coal and is the lowest ranked coal in terms of calorific value (or stored energy) (Ambrose, 2021) and has a high moisture and sulphur content. It would be difficult perhaps to argue that increasing the use of the fossil fuel lignite, with all its associated damage caused by air quality and greenhouse-gas production, is better than the zero-carbon energy production of nuclear power. At the time of writing, Germany is conducting a review of its policy to decommission its nuclear plants due to its desire to reduce reliance on natural gas supplied by Russia in the light of tensions in Europe caused by Russia's invasion of Ukraine.

So from this, we can see that "safe" is dependent on a number of things:

- The perception of the individual.
- The individual's biases.
- The memory of the individual in relation to the same or similar events.
- The individual's levels of:

  o   Skill,
  o   Knowledge,
  o   Aptitude,
  o   Training,
  o   Experience.

- And the cultural, historical, and societal environment in which the context is viewed.

Safety is, therefore, relative. Something is safe *for a given value* of "safe". As we have seen in its dictionary definition, the important phrase is *not likely* to cause harm. And that likelihood is the relevant factor to which we must apply our minds. Turning away from the extreme examples we have discussed and focusing on the products and processes that we create and implement in our modern world, we can apply one more factor that affects safety. That factor is using the product or applying the process *in the manner for which it was created*. This is important because we must understand the environment in which the product or process is to be used and by whom.

In *An Effective Strategy for Safe Design in Engineering and Construction* we discussed how designers of everything from computer software to power stations could ensure that their output (product, process, marketing strategy, or whatever) has safety embedded as an inherent trait from the very outset. In following this strategy, our product should enter service in the best possible shape to prevent harm occurring through its use or in the interaction that people have with it. We discussed the importance of considering such matters as use, maintenance, repair, and disposal, and how these need to be controlled. This control comes in the form of information, instruction, and training, which, when developed in tandem with the product's development, offer the most apposite method of ensuring proper use. What we must examine next is the ongoing control of that use; we must ensure that the information, instruction, and training remain in place, that it remains relevant, and, above all, that it remains adhered to.

2.2 Lots of legal reasons

The core reason for designing things with safety foremost, in the UK and from a work-oriented perspective, is Section 6 of the Health and Safety at Work etc. Act 1974 (HASAWA, 1974). The wording of this section is as follows:

1 It shall be the duty of any person who designs, manufactures, imports or supplies any article for use at work

   a   to ensure, so far as is reasonably practicable, that the article is so designed and constructed as to be safe and without risks to health when properly used;
   b   to carry out or arrange for the carrying out of such testing and examination as may be necessary for the performance of the duty imposed on him by the preceding paragraph;
   c   to take such steps as are necessary to secure that there will be available in connection with the use of the article at work adequate information about the use for which it is designed and has been tested, and about any conditions necessary to ensure that, when put to that use, it will be safe and without risks to health.

This section of legislation quite succinctly deals with the necessity not only to ensure that safety is embedded in the design from the outset but also that thought and attention is given to the repair and maintenance aspects as well as any information, instruction, and training that is required. This clearly requires the designer to be cognizant of the operational environment in which the product is to be used. A fire alarm, for example, is going to be used by persons who will be potentially experiencing some heightened level of stress—assumingly having just discovered a fire—and, therefore, complicated instructions would be undesirable. Similarly, a fire alarm is an important safety-related system, so it may not be considered appropriate to allow just anyone to maintain or repair it. Therefore, not only should we be considering the access to controls and operational components but also any training material for maintenance and repair engineers to use.

There are two important issues that we should discuss with this particular piece of legislation which may explain the seemingly apparent lack of consideration that is often given to it. Firstly, the Health and Safety at Work etc. Act 1974 is very obviously aimed at health and safety. The act came into being partly in response to the tragic incident at Aberfan in 1969, and partly in response to the fractured and arguably haphazard way that work safety was regulated in the UK. Up to that time, workplace safety was prescribed through a number of successive, and in many cases prescriptive, Factories Acts which were first introduced in 1833. The Health and Safety at Work etc. Act 1974 swept away this prescriptive approach and introduced the concept of assessing risk based on individual situations. This is the approach subsequently followed by many legislators around the world. The intention being that the person creating the risk—that is to say, the owner or employer—was best placed to understand that risk and therefore create a system that would manage it.

The term "health and safety" was relatively new at the time the act came into being; up to that point, relevant acts had referred variously to "safety" or "welfare". Today, of course, the term health and safety, and its ideals, has become de rigueur—perhaps even something of a cliché. Certainly, there are those who do not feel well-disposed to consider it as part of their central job function; and designers can be amongst these. The reasons for this are outside the scope of this book, but there is a sense that designers have not engaged with health and safety legislation to any great degree and that health and safety practitioners have similarly not engaged with designers to any great degree either. This has been unhelpful, leading to much rework, unnecessary cost and wastage, and potentially unsafe products and processes. Encompassing health and safety legislation in any design project is as important as encompassing legislation and standards on material types, sizes, power ratings, etc.

The second and equally obvious reason is the inclusion of the "at work" element of the act's title. The designer of an office building may not necessarily contemplate that even though they are designing a place of work that the building design is, indeed, "at work". Similarly, the designer of a new type of kettle for elegant suburban homes may not consider that it will grace the worktop of a workplace canteen. This is not to say that domestic kettles—or, indeed, any other "non-work" item—has no regulation. Up until 1992, products for domestic sale and use in the UK were regulated by British Standards. After 1992 they were regulated by the tortuously complicated New Approach Directive from the European Union, otherwise known as CE Marking. This in turn led to a number of regulations, enabled by the Health and Safety at Work etc. Act (HASAWA, 1974), to come into force in the UK which covered a multitude of work equipment and procedures. Suffice to say that, whether an item is designed for use specifically "at work" or not, it would be extremely churlish to exclude the ideal of safety on the basis that "at work" regulations do not apply. At the time of writing, the CE marking requirements still apply in the UK despite it leaving the European Union, but are to be replaced by the almost equivalent UKCA marking regime by the end of 2022.

For the sake of clarity—and, perhaps, sanity—we shall continue with the assumption that all designed outputs are subject to the regulations identified in this chapter, whether they are work related or not.

The 1974 act was revolutionary in safety legislation terms because it moved away from the prescriptive nature of previous legislation and placed the onus of safety very firmly in the hands of the risk owner (or hazard creator, depending on your point of view). This may seem unimpressively unimaginative today, but it must be regarded in the context of the times. An example of this is the case of Summers (John) & Sons Ltd v Frost where an operator of a grinding machine (Mr Frost) was injured when his thumb contacted with a grinding wheel (1 All ER 870, 1955). The requirements of the Factories Act 1937 were that dangerous parts of machinery should be securely fenced in order to prevent injury and this was the verdict of the original court as well as later appeals. The operator's employers—Summers and Sons—argued that adequate guarding of the grinding wheel had been provided and that to protect the entire circumference of the grinding wheel would render the machine unusable. The original court, as well as subsequent appeals, found that the law's requirement was absolute and, therefore, the situation arose that all grinding machines in the country were, in fact, illegal. This bizarre situation was remedied by the Abrasive Wheels Regulations 1970, which required a modified standard of guarding. These regulations were subsequently replaced in 1992.

The Robens Committee was aware of these inconsistencies and hence created the risk-ownership concept we understand today. This concept is framed in three parts: creating a safe workplace; creating safe work practices, and; providing relevant information, instruction, training, and supervision. Various regulations since 1974 have supported this framework, such as the Workplace (Health, Safety and Welfare) Regulations 1992 (WHSWR, 1992) in connection with the workplace; the Management of Health and Safety at Work Regulations 1999 (MHSWR, 1999) in connection with the assessment of risk at work, and; the Provision and Use of Work Equipment Regulations 1998 (PUWER, 1998) in connection with the provision of information, signage, and so forth for operators of work equipment.

The Provision and Use of Work Equipment Regulations (PUWER, 1998) was the UK's implementation of the European Union's Amending Directive 95/63/EC (95/63/EC, 1995) to the Use of Work Equipment Directive (Directive 2009/104/EC, 2009) which was concerned with harmonizing the way all work equipment was to be used across the EU. PUWER states that the regulations are applicable to "any machinery, appliance, apparatus, tool or installation for use at work (whether exclusively or not)". It may be seen from this that our previous example of a fire alarm sits firmly in the category "installation" as defined by the regulations. The regulations cover matters such as the suitability of work equipment; its maintenance and inspection; information, instruction, and training relevant to its safe operation; the location, layout, and identification of any controls, and; the requirement for protection from dangerous parts. Because PUWER relates to anything and everything we use to perform work they are perhaps the most relevant to us in terms of ensuring continuing safety. And yet, they remain poorly understood and implemented, even amongst some safety professionals, as they are perhaps too often seen as being quite prescriptive.

It is perhaps true that PUWER lacks the nuanced risk assessment approach of other regulations. Instead, there are pertinent questions such as: are the controls functioning; are warning signs in place; is there a maintenance schedule; and so forth. There is also a sense that they are somewhat compliance-driven; having as they do an almost "tick box" fundamental to them. This is not so. PUWER provides an excellent framework for judging many aspects of ensuring the initial safety of all types of work equipment and installations, as well as such peripheral matters as training, instruction, and maintenance. These are all important points to note in the work process and by setting a prescriptive methodology of checking off these various associated items and functions, an overall picture of a workplace, and its wide variety of equipment, will begin to form.

Ensuring the continued safety of structures—workplaces and living spaces—falls under two distinct channels of legislation. For workplaces, the Workplace (Health, Safety and Welfare) Regulations 1992 (WHSWR, 1992) apply, and for living spaces, a myriad of building regulations administered by local authority building control. Of course, there are situations where both of these types of regulation are applied either consecutively or concurrently. Whilst an office block is being built, for example, the construction will have to meet with building regulations. Once it is completed and people start to work there, it will be regulated by the workplace regulations. Conversely, a residential block of flats will contain areas that will need to be accessed by people performing work functions: dustbin stores; lift machine rooms; dry riser pump housings, and so on. These areas must be compliant with the workplace regulations.

In her report, Building a Safer Future (Hackitt, 2018), Dame Judith Hackitt notes that existing frameworks of regulation and standardization can frustrate a cohesive approach to safety, albeit that her report was centred on high-rise residential buildings. In it she notes four principal areas of concern: ignorance of the existing regulations; indifference to delivering quality product; lack of clarity on roles and responsibilities, and; inadequate regulatory oversight. From these four points, we can see that culture plays a large part in these failings. The report's response to the issues it raises is the implementation of new, purpose-led regulation to improve standards across the construction industry, although this has translated into regulation that only refers to "higher-risk" residential buildings above 18 metres in height and regrettably shies away from existing (and all-encompassing) construction regulations. Dame Judith's report was a direct result of the Grenfell Tower tragedy in which 72 people lost their lives, hence the reason for dealing with high-rise buildings. But this unfortunately misses those fatalities in all domestic dwellings in the UK overall, which in the year to June 2021 stood at 249, up from 235 the previous year, with domestic dwelling fires accounting for by far the largest percentage of fires attended by the fire and rescue services (Fire and Rescue Incident Statistics: England, 2021). This shows that there is a deeper issue than that purely concerned with high-rise buildings and points to the need for all relevant agencies, institutes, and professionals to face up to this failure in safety.

And it should not be imagined that the construction industry is alone in facing a crisis of confidence or quality. All sectors of industry have similar issues which are readily relatable to the points raised in the Hackitt report. Some of these issues are:

- Cost
- Design
- Engagement
- Procurement
- Accountability

2.2.1 Cost

The seemingly endless pursuit of cost-saving has brought about a race to the bottom that cannot be justified and has been shown, time and again, to be irrelevant or even counterproductive to overall cost implications. One of the worst offenders for this is, ironically, central and local governments who procure assets through a variety of departments, each with their own budgets. This creates a situation where often the capital expense is pared down during specification at the expense of the operational stage later on. A cost-saving innovation at the construction stage, saving one department money, can subsequently translate into a more expensive asset for a different department to maintain during its lifetime.

2.2.2 Design

Anyone who designs, stipulates, or specifies on any project is a designer, and they should be cognizant of their responsibilities in this regard. This has not always been the case and design often wanders critically from the original intent of the client—a tendency known as scope creep. Clients, too, are culpable here by not understanding the critical importance of delivering a robust statement of requirements to the designer(s) from which the designer can work. A well-informed statement of requirements allows the designer to flex their creative muscles within properly defined tolerances that ensure that the client's ambitions for the design are fully met.

2.2.3 Engagement

Without the full engagement of all stakeholders, both present and future, a project is likely to deliver a product that has not fully considered all aspects of its ownership, use, maintenance, repair, and disposal. Without this consideration, errors and omissions at the production/construction phase are likely to create unsafe situations that someone, at some stage, will have to encounter, manage, and mitigate. This may be the person who at some stage has to replace an awkward valve, demolish a wall of pre-stressed concrete, or fight a fire on the 20th floor.

2.2.4 Procurement

Engaging with the individual or department with responsibility for procurement is vital in any project to ensure that decisions regarding cost centres and accountabilities are properly arrived at. And procurement personnel should be actively engaging in projects to drive this process and to ensure that they are fully aware of all decisions made on a cost basis which could affect levels of safety, either immediately or in the future. There should also be mature conversations centred around the possibilities of spending greater sums on the capital project in order to deliver safer—and therefore less expensive—operational outcomes. Appreciating that the cheapest is rarely the best would be a good start.

2.2.5 Accountability

The Health and Safety at Work etc. Act 1974 was a milestone piece of legislation in that it required employers to account for their own risk and provide suitable solutions to it. This was enhanced by the Management Regulations in 1999 (MHSWR, 1999) which clearly defined these solutions, providing excellent guidance on doing so in the process. The Construction (Design and Management) Regulations 2015 stated, quite plainly, that it was the client in any project—that is to say, the person holding the purse strings—who was ultimately responsible for safety (CDM, 2015). Regrettably, all these pieces of legislation, like so many others like them, tend to be forgotten when it comes to the implementation of safety in so many projects across all industry sectors. This is due to a number of reasons—government apathy, regulator reluctance, public and media perception, and lack of business engagement—but ultimately stems from a perception that "health and safety" is confined solely to safe work practices. It is time that this perception was caused to change.

2.3 Definitions of safety

Amongst those professions which deal with risk and its identification, management, and mitigation, there are subtle but significant differences in some of the terminology. Some professions favour one word over another, for example, or they may use phrases that are uncommon outside of their community. For some terms, there are also accepted differences from standard definitions: the term "hazard", for example, is defined in the Oxford English Dictionary as meaning "a danger or risk", when both these terms have quite different (and specific) alternative meanings in the realms of risk management. In order to alleviate any confusion that might arise, we have listed below some of the words and phrases we use, along with their definitions in the context of this book.

- Hazard
- Risk
- Consequence
- Risk assessment and matrices

- Mitigation
- Criticality
- As low as reasonably practicable
- Safe
- Safety case
- Golden thread

2.3.1 Hazard

The terms hazard and risk are often conflated with one another, and yet in risk assessment and management terms they have very clear definitions. A hazard is something which can cause harm: it is the source from which harm can emanate, whether this is in terms of injury or loss. Examples of a hazard are machinery, electricity, or a chemical process—all of which facilitate the possibility of an injury occurring. Others are financial acquisition, management, and corporate culture, which might all lead to loss. From this, we can see that a hazard does not have to be a physical item, the obvious things we can see and touch and interact with that can cause us harm—and the sort of things that, generally, we concern ourselves with in day-to-day safety. Investing in a new foreign market is a hazard as it exposes the organization to potential financial or even reputational losses. The corporate culture of an organization, when allowed to become toxic, can expose that organization to loss of goodwill, productivity, quality of personnel, and perhaps even longevity in business.

For every hazard there is a degree of variability, of intensity if you like, of how much harm or loss is possible, or how many people or how much property it may affect. The possibility of a harmful outcome from any given hazard, and the scale or impact of that outcome, is called the risk.

2.3.2 Risk

We subconsciously assess risk all the time in virtually everything that we do. Risk is the possible type and size of harmful outcome, or range of outcomes, that *may* emanate from a hazard. The type of outcome—or harmful event—that we identify as being a risk is quantified by two metrics: these are likelihood and severity. How likely is the potentially harmful outcome to occur, and how severe will the effects of that outcome be? For an example, let us turn to that most British of hazards, the weather.

It's late summer, and you are leaving for work in the morning. You look out of the window to see what the weather is doing. It looks a little grey and overcast—conditions that suggest it may be inclement, but you don't recall the weather forecast yesterday saying anything about rain today. You open the front door and inhale: the air doesn't "taste" of rain and feels slightly warm. You decide not to take a coat or an umbrella, and it turns out to be warm, sunny, and dry all day. You even eat your lunch out in the park.

The hazard in our example is of course the weather, and the risk is that we could get soaked if it rains as we have no coat or umbrella. The likelihood and severity of the rainfall is something that we are assessing from a number of information sources, and we are probably making this complicated assessment in a few seconds. We recall the weather forecast from yesterday that said that rain was not due today, but we are historically familiar with the inaccuracy of weather forecasts, having been caught out many times before. We have made a physical examination of the environmental factors involved—the type and density of the clouds, the quality of the air in terms of temperature and humidity, and certain indefinable characteristics that we sense rather than fully understand. We may feel, for example, that we can "taste" rain in the air, sense an approaching storm, or believe that some ailment worsens or improves dependent on changes in the weather.

We therefore have historical data (in the weather forecast); we have environmental data (by examining the physical atmospheric conditions); and we have our personal analysis data (our senses and the knowledge that often, late summer days can start grey and cool and end up warm and sunny). We assess all this information quickly in order to introduce any necessary control measures (the wearing of a coat or the carrying of an umbrella). In the case of this example, we decide not to take a coat and we are pleased because otherwise we would have had to have carried it around all day, thus crinkling the sleeve of our jacket and tiring our arm muscles. The consequences of misinterpreting the assessment of risk would have led to an outcome of harm: either we would have got wet by not taking a coat or umbrella had it subsequently rained, or we could have got a tired arm from carrying a coat or umbrella we never needed.

The importance of historical data in the analysis of risk cannot be overstressed. In our whimsical example here, the historical weather data certainly plays a part in our analysis of the risk of it raining. This data will show perhaps that from January to April it is almost certainly likely to rain, but from June to September, it is less likely to do so. These are, of course, only probabilities, not statements of fact. But this is precisely what an assessment of risk is dealing with, and some tools for analysing risk are based on the *quantitative* values of probabilities occurring, thus allowing for more accurate predictions of any given outcome. But what if we do not know the historical data? If we are in a foreign country we may not know that at 3 pm every day in August it rains torrentially. We may not be aware of the speed of an incoming tide, the effects of changes in altitude, or even speak the language of the sign warning us about the presence of sharks in the water. We often see, in workplace risk assessments, reactions to risks (i.e., the implementation of control measures) that consider the severity of a potential outcome without fully considering the *likelihood*. This is often due to the lack of understanding of the historical data surrounding that particular event occurring, or from the assessor being overwhelmed in their own mind about how dangerous they perceive the risk to be.

In the model of accident causation (an accident being the antithesis of *safe*), A.R. Hale and M. Hale (Hale & Hale, 1970) noted the importance of human factors such as training, experience, skill, and innate ability (see Figure 2.1). Additionally, the model receives input from the "expected information" of past experience and stereotypes. It is this expected information that can be difficult to gauge—more so

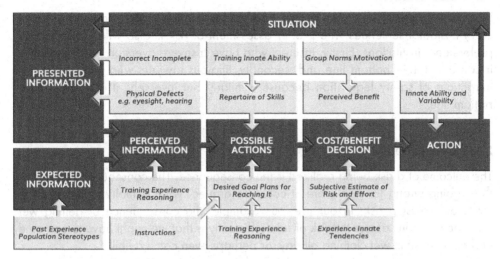

Figure 2.1 Model of accident causation.

than a person's level of training or skill set, for example—because it is often in the mind of the assessor. Few of us might consider entering a burning building, assessing the risk as too great. To a firefighter, however, this would seem well within their acceptable tolerance of risk due to the factors outlined in Hale and Hale's model.

We make all sorts of assessments of risk constantly every day, and, in the main, humans are quite good at it. In fact, our ability to assess risk adequately is one of the principal reasons we have burgeoned as a species. Humans are generally poorly equipped to survive in evolutionary terms: we have poor natural defences; we are not very fast or agile; our sight and hearing are fairly poor; our young are hopelessly defenceless for a long time, and we reproduce at a fairly sedentary rate. This rate of reproduction is low, too, when considered against the rate of maternal mortality during childbirth, a primary reason for which is the astonishing size of a newborn's head in comparison to its body and the mother's birth canal. But it is the size of this head—and, moreover, the brain that it contains—that has brought about mechanical, medical, environmental, and psychosocial techniques that we have created to make up for the fact that humans are, in fact, fairly fragile.

This ability to assess risks in order to stay one step ahead of various predators might make one think that we should be really adept at it when it comes to safety matters. Our natural predators throughout human history, of course, have not always been the ones with big teeth, sharp claws, and low tolerance thresholds. Microscopic pathogens have been responsible for innumerable deaths in humans and it is no small wonder that the humble antibiotic is perhaps the greatest saver of human life ever. In contrast, that other "great" human invention, lead in petrol (gasoline), has been the largest killer. Our assessment of risk, then, is not always up to scratch, and certainly not when there is a pecuniary advantage on offer.

Nor, it might seem, when we are at work and operating under regulations which carry custodial sentences. The proper assessment of risk is also wildly affected by professional institutions. Firstly, insurers who have a severe aversion to risk because it can affect their bottom line and, secondly, litigious lawyers who embrace minor infractions of safety legislation because, similarly but inversely, it can affect their bottom line.

2.3.3 Consequence

The outcome of a risk being realized is the consequence. The consequences of something going wrong (or right!) are, it could be argued, what actually matter to us in terms of assessing risk. In our previous example, the consequence of getting wet because we didn't assess the risk of rain sufficiently is that we might have to go home and change, sit in wet clothing all day, or perhaps even catch a chill.

Often in assessing general workplace risks we forget or ignore the potential consequences. This might be because we have wrapped them up within identifying the risk itself: "If I do not take an umbrella, it might rain and I will get wet". But this can be detrimental to understanding the potentiality of the risk; following it through to its natural and plausible conclusion(s). In complex systems, consequences are used to determine the *criticality* of any given outcome occurring. This, in turn, helps to inform the mitigation process, thereby delivering suitable and sufficient control measures to the right points in the system. A power transformer, for example, might run the risk of failing if the oil coolant level is not maintained. The consequence of this might be that a large proportion of the local population, including a hospital, would be without power. Therefore, control measures relevant to ensuring the oil is correctly measured, recorded, and topped up need to be put in place *commensurate* with this consequence.

2.3.4 Risk assessment and matrices

Aside from assessing risk because we are terrified of either paying higher insurance premiums, being sued, or being incarcerated, there is also a plethora of literature on what a good risk assessment looks like. There are resources on what information should be contained in it; how it should be displayed; how it should be formatted and tabulated, and; how it should be quantified. We learn of adjectives for the likelihood of a harmful event ranging from "impossible" to "certain", and severities ranging from "negligible" to "catastrophic". We see risk assessments that consider the highly unlikely and yet ignore the highly probable, and we sometimes see people referring to risk assessments as a safe system of work. We also see people demanding to see a contractor's risk assessment before starting work on site whilst ignoring the methodology that the contractor actually intends to deploy in order to complete the work. An assessment of risk is, of course, exactly that and is the document that should be used to inform the way that work *should* be done in order to minimize the risks identified.

Additionally, we see risk assessments complicated by the use of a risk matrix and, again, there is plenty of advice available on the best type of matrix to use. The concept of the risk matrix—where likelihood and severity are multiplied together in order to give a relative magnitude of perceived risk outcome—is well understood in areas where some form of quantification of risk is possible or desirable. Chemical processes and engineering are examples where an outcome of a particular event can be quantified in some way, usually in terms of probability. The probabilities of particular events happening, and the predictability of the outcome of those events, leads us to be able to quantify the resultant level of risk. In general day-to-day usage, however, the value of quantifying risk in this way is questionable at best. This is mainly due to the fact that without quantified probabilities—actual numbers with which to make informed calculations—the allocation of numbers to any particular event or its outcome becomes entirely subjective on the part of the individual carrying out the assessment. There is also the issue of the descriptors used for each "level" of likelihood and severity described in the matrix. These can often suffer from rather prosaic language, perhaps describing "multiple fatalities and catastrophic damage" in a risk assessment for office work, for example. This is clearly a hangover from the sort of descriptors that might have been used in, perhaps, a chemical process matrix and calls into question the verisimilitude of the risk assessment as a whole.

2.3.5 Mitigation

In the hierarchy of risk management, the most effective mitigation is that of elimination. By eliminating a risk posed by any particular hazard, it would seem no longer a possibility for a harmful outcome to occur. Would, though, that this was true. It is almost entirely probable that by eliminating one risk we will be introducing another somewhere along the line. An engineering company, for example, which generally works with mild steel as a material may perform a small amount of work on aluminium. Working with aluminium—in particular, when grinding—introduces risks connected with aluminium's explosible properties when ground into small particles. Mild steel, by contrast, does not suffer from this issue. Consequently, our engineering company may decide to contract out their work with aluminium, basing their decision on the cost-benefit analysis between the reduced margin they will be able to command on the work compared to the cost of specialized extraction equipment which minimizes the risk of aluminium dust igniting.

This may seem a sensible solution and is the sort of transfer of risk that can be seen across virtually all forms of business: the specialization of companies who deal with particular types of risk in order that other businesses do not have to invest in safety-related equipment and training, and all the resultant maintenance costs and insurances and so forth. And as elimination is the most effective risk management strategy, this all makes perfect sense, except of course for those businesses that do perform this specialized sort of work who have to invest appropriately. But if the global pandemic of 2020/21 taught us anything, it is that this reliance on supply chains can be extremely fragile, especially where pan-continental shipping is involved. By eliminating the risks from dealing with a particular safety-related issue, we invariably introduce a different type of risk that is connected with issues regarding reputation, quality, finances, or reliability of the supply chain. Our engineering company may have removed a safety risk and cost burden to themselves, but they have introduced the risk that if their supplier goes bust, they might lose the ability to fulfil their customers' orders. What must be considered is that sometimes we can only eliminate part of the risk. As in the example above, we have removed the risk of working with a potentially explosible material, but we have not eliminated the risk of working with metals in toto.

Substitution, the next level in the hierarchy of risk management, is also subject to risk replacement rather than risk removal. Often used in chemical processing, the premise of substitution is to replace a dangerous or potentially dangerous substance or process with one that is less dangerous (or safer, depending on your preference for terminology). A global example of substitution was the replacement of petrol-engine cars with diesel engine ones throughout the late 20th and early 21st centuries. Diesel engines had been for many years considered inferior to petrol for performance reasons: they generally cannot rotate as quickly; they are harder to start in cold weather; the fuel itself becomes waxy in low temperatures; and they emitted a smoky exhaust.

Various developments, however, began to see diesel as a viable alternative. For example, injection systems were improved; fuel additives were developed to reduce the waxing point of diesel oil; engine component manufacturing improved; catalysts were added to exhaust systems; and gearboxes with more flexible gearing options became more prevalent.

The general benefits of diesel power over petrol—better fuel economy, longer times between servicing, increased longevity, and so forth—became, therefore, more attractive. Unfortunately, diesel had an issue that seemingly took several decades to address properly. Particulates generated by the combustion process were causing alarming levels of health issues in cities all over the world. We had traded the harmful effects of exhaust fumes from petrol engines—in particular those containing lead—for an epidemic of asthma and other breathing-related disorders caused by particulates. The usual knee-jerk reactions of government were complemented by initiatives to reduce particulate emissions, like the Euro engine designations introduced by the European Union in 1992, which culminated in the Euro 6 standard requiring the use of an aqueous urea solution to all but eliminate particulates. The upshot of all of this is that measures to substitute particular risks can often generate subsequent risks with similar, or possibly even worse, consequences, just as is possible with the strategy of elimination.

As with elimination, we can see that substitution is often only part of the solution; and indeed, it is sometimes only possible to substitute *part* of the risk. It is the cumulative effect of this "chiselling away" at risk that provides the most appropriate solution: that which is defined in health and safety law as *reasonably practicable* (HASAWA, 1974). In Figure 2.2, we see this effect of progressively reducing risk—represented here by the solid ball—by using consecutive and complimentary methods.

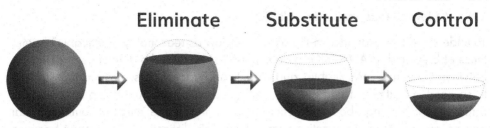

Figure 2.2 Hierarchy of risk management.

Control measures—or mitigation—are often seen as risk *reduction* techniques, whereas elimination or substitution are perhaps considered more as risk *removal* techniques. This is not correct. Eliminating or substituting the risk from a particular hazard may well introduce risks from other existing hazards, or from newly introduced hazards, as in the example of supply chain issues. All forms of risk mitigation should therefore be considered equivalently; that is, that risk cannot be completely removed unless there are no hazards from which risk emanates. Clearly, though, a world without hazards would not be one with a great deal happening in it.

2.3.6 Criticality

In the management of risk, criticality is an indicator as to where in the list of priorities any given outcome, or consequence, of a risk being realized sits. In the guidance to the MHSWR (1999) (HSG65, 2013), the Health and Safety Executive (HSE) refer to "risk profiling," which engenders these four main points:

- The nature and level of the threats faced by an organization.
- The likelihood of adverse effects occurring.
- The level of disruption and costs associated with each type of risk.
- The effectiveness of controls in place to manage those risks.

The consequence of the risk is contained within the determination of the "level of disruption and costs associated" with any given risk being realized. This, too, should provide the levels of criticality that one should consider.

Criticality is used in analysing complex systems and maintenance scheduling. It is also considered in risk registers, although usually in terms of the cost implications. This is still a measure, though, of the amount of effort that we might need to apply to each risk; that is to say, a measure of prioritization due to the critical nature of the consequences. An example of the relevant criticality of, say, the indication of speed would be that of a car and an aircraft. If a car's speedometer fails due to perhaps the drive dog from the gearbox wearing out or breaking, it is unlikely to prove catastrophic. A driver can judge their speed against other traffic or the environment and, in any event, can pull into a garage to have the repair made. If the sensor for an air speed indicator of an aircraft in flight fails, especially if there is no back up sensor and it is inextricably linked to the autopilot, there is a real chance of a dire consequence.

2.3.7 *As low as reasonably practicable*

To understand the resonance of the phrase "as low as reasonably practicable" (sometimes abbreviated as ALARP) it is perhaps best to start with the difference of definition between the words "practicable" and "practical". Practical is something that is likely to succeed in real circumstances or is concerned with the actual doing of something rather than theorizing about it. Practicable is a narrower definition of something that *can* be put into effect or practice successfully. This distinction is important because something that is *practicable* may not necessarily be *practical*.

The word practicability is used widely in English law and indeed was so long before the Health and Safety at Work etc. Act 1974, where it became the cornerstone of the application of health and safety law. The legal definition of practicability came in the case of Edwards v. National Coal Board (1 All ER 743, 1949), where the Court of Appeal gave this determination:

> *"Reasonably practicable" is a narrower term than "physically possible" ... a computation must be made by the owner in which the quantum of risk is placed on one scale and the sacrifice involved in the measures necessary for averting the risk (whether in money, time or trouble) is placed in the other, and that, if it be shown that there is a gross disproportion between them—the risk being insignificant in relation to the sacrifice—the defendants discharge the onus on them.*

It can be seen from this determination that the requirement of health and safety legislation to reduce risk to a level *as low as reasonably practicable* is based on the efficacy of an initial assessment of risk and its probable consequences. The greater the likelihood or severity of risk—that is to say, the greater the harm of any potential outcome—then the more extensive the mitigation or control measures must be. The converse is also true, and the Health and Safety Executive give these two extremes as examples (HSE, n.d.):

- To spend £1m to prevent five staff suffering bruised knees is obviously grossly disproportionate; but
- To spend £1m to prevent a major explosion capable of killing 150 people is obviously proportionate.

It is also worth noting that the phrase "so far as is reasonably practicable" (sometimes abbreviated as SFARP) is more often used in health and safety legislation, whereas "as low as reasonably practicable" is more often used by those involved in risk management or engineering, for example. The two phrases are to all intents and purposes, however, interchangeable.

2.3.8 *Safe*

In Section 2.1, *What does it mean* we discussed the concept of safety and established that there are many aspects which can affect the determinable level of safety in any given situation. In essence, something is safe when *it is not likely to cause harm* and, as we noted previously, when *used in accordance with its intentions* and *within its operational parameters*. We may consider, as an example, a hammer. This may be perfectly well designed and well made; able to cope with a lifetime of striking a variety of materials and fixings for which it was developed. It may have a perfectly balanced shaft with a slip-resistant grip and a sealed cleat to prevent the head ever loosening. Striking the wrong material, such as an iron anvil or a human thumb may well, however, cause harm; either to the hammer or the hammer's operator. We can deduce from this, then, that in order to use anything safely, we must have designed it for the purpose for which it is to be used and provided suitable instruction and training for that purpose.

Specifying precisely what we intend to use any particular product for is the reason why everything should be designed according to a statement of requirements. And this is why we have hammers of varying sizes and styles, and made of various materials. Instructions for use may be long and complex—and may involve or influence subsequent training regimes—or they may simply consist of indications of limitations of use. This is why even something as innocuous as a stepladder comes with a label showing the maximum load capacity; that is to say, the load criterion it was designed to meet.

2.3.9 *Safety case*

The justification that any product is safe to operate is established by the design risk assessment(s) carried out prior to production. This assessment may be for the entire product—where it is a single or simple one—or there may be many assessments where the product is made up of multiple components or is particularly complex. Clearly where there are multiple components, the risk assessments must align at some stage to ensure that the overall assessment of risk—and therefore the operational parameters—is cohesive. Two components or materials may, in themselves, be perfectly safe to use and operate, but if in combination they produce an additional hazard, this would be detrimental.

To qualify all this data, and to consider the *overall* operational parameters (as well as any operational limitations), there would be a safety case written up. This is the body of evidence that identifies the hazards and risks associated with the product. Any operational mitigations that are required can be defined in the safety case in order that safe and effective operational procedures can be developed in line with the risks identified. This is the transition from "safe to operate"—as curated by the design process—to "operated safely". This operated safely stage is of course where standard health and safety protocols and regulatory compliance take over to ensure continued safe operation.

In essence, the health and safety file or technical file required by specific types of safety legislation forms part of the evidence required to substantiate a safety case. These files should demonstrate that any relevant risks have been addressed, and to what extent. There is unfortunately much disinclination towards properly addressing the production of a cohesive health and safety file for any given project, not least in the construction industry where it is a legal requirement. This may be due to the inclusion of the words "health" and "safety" in that order, which many outside of the safety profession regard with unease, if not suspicion. Engineers also, who frequently use safety cases to demonstrate the safety of their creations, sometimes fail to note the importance of this type of information in support of their case. The quintessential similarities of all three—safety case, health and safety file, and technical file—are that they are bodies of evidence which demonstrate the reasons certain decisions were made affecting the safety of the product and what measures are in place to sustain it. With a safety case, this evidence is backed up by arguments demonstrating that the evidence supports the claim (that the product is safe), whereas with reference to the other two, these arguments are missing and the documents merely state the required measures of control. The fact that these documents exist (or, at least, should exist) is by implication part of the argument that the evidence is present.

2.3.10 Golden thread

The concept of having a system for recording all information concerning the design, production, and safe operation of a building has gained traction in recent years, and in particular since the publication of Building a Safer Future—Proposals for reform of the building safety regulatory system report (MHCLG, 2018). Although in response to what may be considered more "traditional" construction issues, and in particular in response to the Grenfell Tower disaster, it can be seen that to understand how a design interacts with users and operators throughout its life is a fundamentally essential thing. This was, of course, always the intention of the health and safety file introduced in construction regulations dating back to 1994 but which has, regrettably, never really been given the credence it deserved. Having functional safety information fully supports the requirements of CDM in the production of a health and safety file, and can also be used as substantiating evidence in support of a safety case or a technical file.

The ambitions of these various methodologies for presenting relevant safety information should be considered equivalent, whether they have been developed for design or engineering purposes or solely for regulatory compliance. The object is to encourage the consideration of safety as an embedded feature of design; provide the user/operator with evidence of that consideration, and; inform the user/operator of risk factors that may require further control measures. Additionally, it should be borne in mind that complying with both the letter and *spirit* of the construction regulations will ensure that, to a greater extent, the contents and ambitions of each type of methodology are inherently provided for. This is the reasoning we examined in our book *An Effective Strategy for Safe Design in Construction and Engineering* (England & Painting, 2022).

The creation and regular reviewing and updating of a risk register during the design and construction/production phases helps to ensure that the management of risk is maintained. This also prepares and develops the "thread" of information necessary for the health and safety file, technical file, or safety case. However the end user is informed of any residual risks in the product, the purpose of the "thread" is to inform them of any safety information for those who have to operate, maintain, repair, or dispose of the product, and the relevance of creating and disseminating this information properly should not be underestimated. The engineering profession has developed and operated safety cases for many years: they are a principal component of how engineers demonstrate how their ideas have been designed and made safely and how this is then translated into safe operation throughout the life span of the product.

The reasons and tools for creating a thread or sphere of safety for our designed products have long been in existence. It has also been our legal duty to ensure the safety in the design and manufacture of them for many years—at the very least since 1974. It is also our moral and ethical duty to do so. The introduction of yet more initiatives and legislation to reinforce what is already painfully apparent may seem a little superfluous, but anything that helps to prevent error and promote safety cannot be a wrong move. So long as, however, these new initiatives and pieces of legislation are adhered to and not ignored, just as the previous ones appear to have been.

2.4 Similarities—construction and engineering

Oh, East is East and West is West, and never the twain shall meet.
(The Ballard of East and West, Rudyard Kipling)

Kipling's famous quote is often alluded to refer to the writer's possibly colonialist racist views. But, of course, as is often the case in such matters, protagonists do not bother to read the whole text, as only two lines after this commonly cited quote, the writer states, "But there is neither East nor West, Border, nor Breed, nor Birth". The two men depicted in Kipling's piece may be from geographical locations separated by a great distance—and therefore unlikely to meet in normal events—but their essential characteristics of honour, courage, and mutual respect are the same. They have widely differing backgrounds and yet are essentially similar people. An allusion to the fact that, in reality, irrespective of where a person is from, what they look like, or how they dress or speak, humans are in fact all the same.

The same could be said of the disciplines of construction and engineering. Neither are usually considered to be the other and, in order to resolve any grey areas, we have even derived the term "civil engineering" to establish a peacekeeper of sorts. And yet, engineers are involved in construction all the time: mechanical engineers, electrical engineers, structural engineers, acoustic engineers, safety engineers, fire engineers, and, yes, civil engineers. But construction is seldom considered as engineering and engineers possibly wouldn't consider themselves as constructors. But if an engineer doesn't construct something, does that make them simply a designer? Is that what engineering is: just calculating and designing without actually doing the clever part of turning a drawing into an actual thing? A toolmaker who can take a requirement for a tool and, after designing it, can turn and work a piece of metal into a finished product is undoubtably an engineer. Does that mean a carpenter is also an engineer? Or a bricklayer or landscape gardener? Is engineering actually the skill of developing something from nothing using the skills and materials with which one has been trained and has experience of? The word engineer is an extension of "engine", and yet we have chemical engineers who have never so much as fitted a spark plug.

The Industrial Revolution of the 1700s in Britain marked a turning point in the production and consumption of engineering as we perhaps understand it today. It was, of course, the first and only revolution of its kind—all so called "industrial revolutions" since then are, in fact, *evolutions* in terms of the progress they have added. Simply powering machines by electricity rather than steam or digitizing information that was once written down does not represent the paradigm shift—the complete societal, socio-economic, and industrial upheaval—that the Industrial Revolution ushered in. But even so, metal had been worked for thousands of years prior to this time—iron, bronze, copper, tin, lead, and so forth. Geography has always played a part in what humans have been able to utilize as engineering solutions. The ancient Egyptians, for example, had plenty of stone with which to build their pyramids but only copper with which to make the tools to work the stone. In Japan, the abundance of wood, combined with the need to make buildings proven against earthquakes, meant that carpenters developed advanced yet elegant types of joints to resist collapse. Whatever the need, humans have an innate ability to engineer solutions to problems using available materials, existing knowledge, and creative thinking. Hence our ability, despite all appearances, to elude our natural and ravenous predators whom we discussed earlier.

The upshot of all of this is that a solution to a problem requires knowledge and expertise. And although both a carpenter and a blacksmith could probably solve the same problem using the materials with which each is most familiar, we might not generally consider either of them to be "engineers". And yet, engineering is solving problems and construction is building the solution. Both are equally important and neither can exist without the other. Sometimes they exist within the same person, and sometimes it takes a team of people with individual skills to create the final product.

Within any discipline, there are always engineering principles that need to be adhered to. These are generally connected with the physical properties of whatever materials one happens to be working with. A civil engineer, for example, understands the tensile properties of concrete and how this can be affected by the addition of other materials such as plasticizers, glass fibres, steel rods, and so forth. Similarly, a carpenter must understand the properties of various woods; how they work with each other and with other materials, such as iron or brass fixings. They must also understand the inherent strengths of various joint types and which woods are most able to be worked to fine tolerances. Anyone who does not consider carpentry to be engineering would do well to study the eminently graceful and technically exquisite wooden structures of Japanese carpentry; held together without screws, nails, or other modern fixings.

The division between construction and engineering (and possibly what some might rather demeaningly refer to as "trade") may be due—and certainly perpetuated—by the divisions between the institutions and societies which represent the various disciplines involved. There is always a certain amount of introspective thought amongst these types of groups, in whatever walk of life we care to look into, and we must always be aware of the potential for "silo thinking". This is where individuals or groups of individuals can become mired in their own aspect of a project to the detriment of communicating or understanding other aspects. This can be due to cognitive biases such as confirmation bias, or "groupthink". It is a well-recognized attribute that some are trying to tackle through various initiatives in various disciplines and industries. It can be countered by proper and effective use of the "four Cs": communication, cooperation, control, and competence. We examined in our previous book how even small errors can be amplified many times over as a project progresses, ultimately leading to situations where large amounts of money have to be spent on repairs, alterations, and remediations. We discussed how investing relatively modest amounts of time and money early on in a project can reap huge savings later on. The issue is always persuading those in charge of the purse strings that a penny spent is very often a pound saved, although it is amusing to note how many individuals in charge of often huge budgets have very little concept of this fundamental economic reality.

There have been initiatives set up, for example, by individuals with a keen understanding of the problems of waste in construction and their passion for "getting it right". But waste is not confined simply to construction: the almost farcical (and certainly furtive) remediation that had to be conducted on 601 Lexington Avenue, New York, in 1978 (Werner, 2014), just a year after the building was completed, was as a direct result of failures by the engineers to understand wind-loadings on the building; a point that was picked up by an engineering student studying the design after its completion. This example of getting it wrong, rather than getting it right, is unhappily not unusual, and any initiative to *get it right* should welcome the whole fraternity of individuals and groups who are involved in creating our built world, regardless of howsoever that transpires.

Humans are hugely social creatures, even if perhaps not always *sociable*. We are destined to seek out others who are similar to us and then to generate cognitive biases which exclude others who are not in our "gang". This is true at all levels of society and exists equally in professional doctrines. Surgeons, for example, take pride in the fact that they revert to Mr or Mrs rather than Dr, a throwback from a time when the training for doctors and surgeons followed quite separate paths and surgeons were not required to train as doctors first. Surgeons are also regulated by the *Royal* College of Surgeons, whereas doctors by the *General* Medical Council. Small differences in terminology have been the mainstay of division throughout human history.

But is this division right? Is it helpful? Is there not an argument that cohesive dialogue between disciplines is the correct way forward; to improve our world and create better and safer things? Is there not a need to consider that, irrespective of whether a job title has protected status or not, everyone involved in the creation of a product has a part to play in ensuring that it is the best and safest it can be; not just now but throughout its entire lifespan?

Our carpenter, for example, may be extremely skilled at designing and making things from wood; beautiful, elegant, and practical things. They may, however, not have an encyclopaedic knowledge of different woods and their properties and may therefore engage the help of someone who has. Someone who, perhaps, does not have the artisanal skill to create objects from wood but knows of every type of wood and their individual properties. The carpenter could describe their design and its purpose and work with the wood expert on which wood would be best for the carcass, the panels, and the bracing. The expert might suggest alternatives that cause the carpenter to alter their design slightly to accommodate novel properties of a particular wood that they had hitherto not considered. The end product will be better *in all respects* because of this collaboration.

The Construction (Design and Management) Regulations 2015 clearly state that the regulations are applicable to "any building, civil engineering or engineering construction work" (CDM, 2015). They go on to include construction, alteration, conversion, commissioning, upkeep, maintenance, preparation, assembly, and removal amongst others. Indeed, upon reading the full definition of "construction" in the regulations, and the applicability of them, it is quite difficult to imagine to what the regulations *do not apply*, rather than to what they do. We demonstrated in *An Effective Strategy for Safe Design in Engineering and Construction* that the CDM regulations make an ideal framework for ensuring the appropriate amounts of time, effort, and resources were directed at any design project, regardless of the desired output. Construction, engineering, software, space exploration: whatever the project is engaged in, correctly assessing the potential outcomes from all and every aspect of the project is vital to ensuring the end product is the best and safest it can be. An additional advantage is that, in the UK at least, by following the requirements of CDM, one is also assured of a high level of compliance with regulatory requirements.

Ultimately, the creation of *any* product follows the same simple process: requirement; design; production. The better the integration of all and every discipline involved in this process the better. And, of course, subsequent to the production of the item, it must then be operated, maintained, repaired, modified, and ultimately disposed of. Again, the integration of all and every discipline and profession involved in this subsequent part of its life cycle is crucial to ensuring its ongoing safe operation.

2.5 Why is it important to consider safe to operate and operated safely?

The use of any product obviously begins once the design and production phases are complete, and it is passed to the end user. Throughout the design phase, the actual production safety will have been considered: what materials are to be used and how; what energy systems are required; how large or heavy components will be handled, and so forth. But the design should also have considered the use of the product and, as importantly, the environment in which it will be used. Equally, who will be using the product—experts or a generality of people—and who will be maintaining the product and how? Safety does not start and end with the design phase, and nor does the expectation of safety regulations.

Starting a design with a cohesive understanding of what is truly required of any product is critical to ensuring its ongoing safety throughout its lifetime. The key points of understanding are:

- The client's requirements.
- Market expectations.
- Operational feedback from any previous similar product.
- The environment in which it is intended to be used and any environment in which it cannot be used.
- What regular repairs or maintenance are anticipated, and how these will be achieved.
- Any regulatory or market restrictions that affect its operation, construct, or maintenance requirements.
- How the product will be disposed of, either through reuse, recycling, dismantling, or demolition.

Imagine we are designing a heavy goods vehicle for use in developing countries. The client, a small but prominent commercial vehicle manufacturer, wants to dominate the market for financial reasons but also wants to provide a quality vehicle that will help in the transport of vital goods in poorer countries in line with its corporate social responsibility. Some of the initial considerations for the design might be those in Table 2.1.

Table 2.1 Example design solutions

Consideration	Possible design solution(s)
There are few made-up roads	Develop four-wheel drive capability
	Provide for increased wheel travel in the suspension
	Provide enhanced air filtration systems with readily accessible filters
Repair agents may be few and far between	Ensure mechanical equipment is readily accessible
	Minimize the use of complex or novel systems
	Use standardized fixings to reduce the need for specialist tools
	Reduce the number of different sizes of fixings used
Good quality fuel may be in short supply	Provide enhanced fuel filtration systems with readily accessible filters
	Reduce the unladen weight of the vehicle to enhance fuel economy
Loads may be many and varied	Provide numerous fixing points on the chassis to allow flexibility in body attachments
	Upgrade suspension and brake components

You may be able to think of many more. What is important is that all of these considerations are interrelated. They consider both the *mode* and *environment* of operation: who will operate it and how. The product is designed to be safe in consideration of its use, according to the specifications given to the designers. The resultant product will then perhaps undergo testing, the function of which is to ensure that it is capable of standing up to the use for which it was designed. Testing criteria will be influenced by the design specifications and the user's specified requirements. In use, the product will need to be assessed for risks to the user; risks that, of course, should have been *eliminated, reduced,* or *controlled* at the design stage to a level as low as reasonably practicable. Again, knowledge of the operational parameters and expectations of the product will help the designers implement effective controls at the design stage in order to minimize the risks. Operational limitations on the product, put in place to prevent its exposure to any particular failure, should then be accounted for in operational procedures which, in turn, will inform any necessary training regime. This is what we referred to as safe to operate and operated safely, and applies equally to physical products as well as work processes or systems.

This example also demonstrates a dichotomy in design, and that is the over-designing of products just for the sake of it. We are considering here what a manufacturer in a developed country might do to release a product in a developing nation, and it appears in many ways to revolve around backwards engineering. If we looked at the history of our imaginary vehicle manufacturer, however, we might find that several decades ago, they were making vehicles along these very same lines for their own domestic market. The Land Rover is a case in point here. Developed in 1948 along the lines of the Willys Jeep, the car manufacturer Rover wanted a vehicle that farmers could use both on and off road. It had to be robust, lightweight, adaptable, easy to maintain, and demonstrate dependable performance on a variety of surfaces. Very similar qualities to those we have discussed for our example goods vehicle. The original Land Rover (later known as the Land Rover Defender to distinguish it from other models such as the Range Rover and Discovery) combined these qualities so well that its production run lasted uninterrupted for 66 years with over 2,000,000 vehicles sold worldwide.

Even the Range Rover, launched in 1971, which was to be the luxury variant of the Land Rover, was, essentially, a similarly utilitarian vehicle but with cloth seats and no drain plug in the floor (early Land Rovers could be hosed out inside to remove the accumulated mud from passengers' boots). It was equally adaptable and robust and could be maintained relatively easily. The latest models of both these iconic vehicles are a far cry from their ancestors and feature all manner of electronic systems to aid the driver in traversing the most brutal terrain, but at what cost? It is unlikely that a modern Land Rover could be repaired halfway up a mountain in a blizzard with some scraps found under the passenger seat. (This may not have been possible anyway, but the authors do have first-hand experience, apropos of this type of situation, of repairing the exhaust of a Jaguar XJ at the roadside with a piece of wire and an empty drink can, the electric windows of a Renault with a piece of cigarette paper, and the fan sensor on an FSO with a paperclip. They have since invested in far more reliable cars.)

There is perhaps little doubt that in terms of customer experience, the latest Land Rovers are far superior, with their advanced electronic aids that simplify the user's experience of off-road capability. But electronics are notoriously hard to repair without the proper equipment, which may be hard to come by in the sort of locations that vehicles like a Land Rover might find itself. This may well be a case of over-design, which, whilst improving the safety of the user in extreme driving conditions, actually overlooks one of the principal environmental challenges that the design might face. Similarly, military hardware is renowned for being extremely rugged and is often over-engineered for obvious reason. But short production runs and over-engineering come at a high cost which, although possible for national governments, is less possible for struggling private businesses.

Keeping things simple is not of course confined to vehicle design. The Supermarine Spitfire, the iconic fighter aircraft of the Second World War, was renowned for being an exceptional aircraft to fly and was robust enough to be able to carry on flying even after taking considerable damage. And yet, due to the shortages of materials at the time, the Spitfire was made using wood, canvas, cardboard, and even string in its manufacture. Fast forward to modern-day tactical air combat and we discover the F-35, a plane that not only costs some 30,000 times more than a Spitfire did in its day but is also completely aerodynamically unstable. The Spitfire may have been able to fly with holes in its fuselage, but the F-35 cannot even stay airborne without an array of computers making thousands of calculations a second. Not that that is to suggest that the Spitfire is the better aircraft, or that we shouldn't be looking to technological advancement. The F-35 has been crafted over a long period of time with the benefit of protracted testing; the Spitfire was developed, at pace, during a time of national crisis and supply shortages. It was, in fact, this lack of materials and urgency that forced the design to adapt and improve from a purely functional and operational viewpoint. A case in point is that of the most widely produced gun in the world, the Kalashnikov AK-47, a weapon of great simplicity and yet capable of operating in virtually any theatre of war or environmental condition imaginable.

When the American military wanted to replicate the perceived virtues of the AK-47, their very considerable spending power developed the M16, which, although lighter and more powerful than the AK-47, was considerably less robust and reliable. One of the reasons for this was that the M16 was made to much finer tolerances which in turn led to it being far more prone to jamming up when exposed to harsh operating conditions, such as those in the jungle or desert. The secret to the AK-47 is that it is simple to build, simple to operate, and relatively sloppy in its design tolerances allowing it to deal with the ingress of dirt more readily. It is not a better weapon than, say, the M16 or any other, but it is a *better-designed* weapon for the operational environment in which it is intended to be used.

Although clearly Mikhail Kalashnikov never organized customer focus groups with bands of terrorists and miscreants to discuss the key desirables for an automatic rifle, he would have had the harsh Russian weather as a reminder of what real-world personal weaponry must endure. Hence, the weather in this case was a dominant driving force in the design.

If we consider the word "safe" to mean "without risk" as opposed to "without harm", then we can see that the development of various technology over the years may well have actually gone backwards in terms of being safer; that is to say, exposing the user to less risk. Having a vehicle that does not rely on the driver for descending vertiginous inclines could be said to be safer, or less risky, to the driver; but if that same vehicle cannot be restarted in the depths of the jungle without specialist equipment, then we could argue that it is actually *less* safe, in terms of the function it is meant to perform. The consideration here is that the ability to drive safely and competently off road is merely a matter of training, whereas the ability to perform technical repairs is a matter of training *and* access to the right tools. This important distinction affects the cost/benefit profile of adopting advanced technology over inculcating the user. Similarly, an aircraft that is as aerodynamic as a house brick without its computers working is unlikely to survive long in battle if the enemy discovers a way of defeating those computers. Or, as is so often the case with this form of technology, the computers end up defeating themselves.

Ensuring something is risk free in the design phase is, as we have established, wholly connected with the importance of ensuring that that design is as risk-free as possible during its *entire life* of *intended use*. And only by assessing the risks of its intended use will we be able to provide the best possible design solution.

3   Sources of risk

3.1 Setting out

Anything that we do, from sharpening a pencil to flying a combat jet in a dogfight, has inherent risk—risk that may be considered either positive or negative in outcome, that is to say, either as an opportunity or a threat. An opportunity from a good, sharp pencil is that the line you draw on a workpiece will be more accurate, which will lead to a better quality of fit when you make a cut along that line with a saw. A threat might be accidentally stabbing yourself with the point. An opportunity from flying in a dogfight could be the winning of a battle against the enemy, paving the way to a successful outcome in later battles and, possibly, the entire war. The threats are fairly obvious to imagine.

Risk management is merely the function of identifying the abundant opportunities and threats in any given operation and accurately assessing their potential likelihood and outcome. This can essentially be done by anyone who fundamentally understands the operation in question, or is at least able to ask pertinent questions of those who actually perform the operation. It could also be performed by mathematicians, statisticians, safety advisers, or engineers. It may even require the services of a professional risk manager. There is, however, a flaw to this idea that just about anyone can assess risk, and it is a huge, mud-soaked, raging African bull elephant in the room. The flaw is that everyone, regardless of their experience, skill set, or discipline, will be imbued with their own biases, opinions, historical experiences, and prejudices that will colour their analytical judgement, either positively or negatively. A toolmaker who has worked a lifetime in engineering workshops may be insensible to the risks posed by long-term contact with cutting oil or welding fume, whereas someone who has never set foot in such a workshop may perceive myriad risks that are simply part and parcel of the work of a toolmaker—risks that are mitigated by their high degree of training.

For the benefit of clarity, we shall continue to discuss risk in terms of the threats posed rather than the opportunities. The reasons for this are two-fold: firstly, this is a book about safety and the safe use of things, so threats are of more importance to us; secondly, the proper management of opportunities is often associated with disciplines outside of the scope of this book—for example, marketing, procurement, and finance.

Designing-out safety issues is often a process that involves input from several disciplines associated with the design—the operators, the maintainers, the repairers, the managers, and the safety advisers. The outcome of not adequately considering all the possible factors is that the end design could be less safe than that which it possibly

DOI: 10.1201/9781003296928-3

replaces, or that it introduces hazards or risks supplemental to those considered residual in the design. The authors are reminded of being called in after a flood in the basement of a high-rise block of flats. The water and fire pumps had, correctly, been mounted on plinths to raise them above the potential water line; but the 230V socket for a mains-powered pump—to be used in the event that the building's own pumps failed—was mounted on the wall *below* the potential water line. A prospect that the fire brigade was rightly concerned about upon discovering it.

How could such a thing be missed? Clearly the architect had put thought into raising the level of the pumps in the basement in case of flood, but had missed this one small detail. The building had been built under the auspices of the 2007 incarnation of the Construction (Design and Management) Regulations which meant there would have been a CDM Coordinator (CDM-C) involved. But CDM-Cs were generally more interested in the health and safety of the *construction* rather than the *design*. The 2015 revision of the Regulations changed the role of CDM-C to that of principal designer, with specific responsibility to ensure the application of safety in the design, but invariably clients are concerned with additional costs rather than assiduous design reviews, and often this role is taken on by the architect anyway. The marking of one's own homework rarely leads to the best possible outcome.

3.2 Developing

The site manager during the construction of this block of flats would have been too preoccupied with ensuring the job was finished on time to raise any issues, this being a side effect of the construction industry as a whole not being overly profitable and the spectre of clawback charges should the project overrun is something that keeps construction companies awake at night. The contractors on the ground installing the socket are unlikely to have said anything because raising issues on site is still not seen, even in this day and age, as being wholly conducive to one's career. And besides, if the design drawing says that is where something goes, it must be right. Right? The maintenance engineers wouldn't have said anything because every time they tested the socket, it undoubtedly worked perfectly and, until the day the building actually flooded, they had never witnessed the basement with water all over the floor. Essentially, like the Swiss cheese model used to describe James Reason's view of safety failings, all the holes in the various levels of prevention between concept to operation lined up perfectly to leave a potentially fatal flaw in the product (see Figure 3.1). A tiny change on a computer drawing would have eliminated this problem.

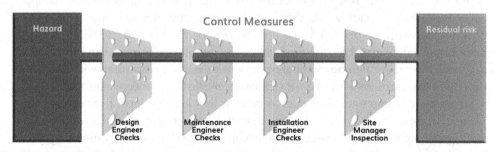

Figure 3.1 Swiss cheese model showing example control measures.

For the benefit of the technically minded reader and to provide clarity for those who may not know: yes, the socket *was* protected by a residual current device, and yes, firefighters do wear heavy rubber boots to protect them from errant current. Though from a safety perspective, that is hardly the point. And in all fairness to the architects, the building was discovered to have a number of other faults which only became apparent when the fire brigade tried to evacuate people in the pitch dark. (Although the fire pumps were mounted above the potential water line, the power to them was not. Had there been a fire, the brigade would have had no water in the dry riser).

This relatively simple issue could have been detected—and, indeed, *should* have been detected—at any number of stages throughout the design and subsequent construction. Consider the following stages:

- Sequential independent design reviews throughout the design phase.
- A full design review prior to the design being signed off.
- The involvement of operational personnel at critical stages of the design:

  o Facilities personnel for the cleaning cupboards;
  o Maintenance personnel for the plant room;
  o Lift engineers for the lift shaft and plant room;
  o Firefighters for the fire resilience, and so on.

- Full discussions between the designers and the contractors prior to the build phase.
- Effective health and safety monitoring during the construction phase that considers what is actually being constructed rather than how it is constructed.
- Proactive monitoring of the build by the site manager.
- Proactive monitoring of the build and the site manager by the main contractor.
- A receptive culture to potential issues raised by site personnel.
- Independent safety audits carried out throughout the construction phase.

The relationship between design and use can be seen in Figure 3.2.

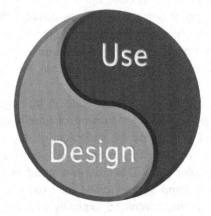

Figure 3.2 Relationship between design and use.

Design affects how something is ultimately used, and the way something is used will affect the design of similar products in the future. A modern aircraft is a highly complex machine that relies on very specialized design. Consequently, it requires many hours of training for someone to be competent to fly it. A fire exit, however, should not require any training whatsoever as the people who will be using it are likely to be: firstly, untrained in its use; secondly, potentially unfamiliar with their surroundings, and; thirdly, in a panic to escape. This is why fire drills are best conducted without people's prior knowledge.

The safety inherent in the design of anything is entirely dependent on how and where it is to be used and by whom. The use of a carpenter's turning lathe must allow the material to be exposed to allow the carpenter to shape it. This is a potentially dangerous activity which requires training—just like the training required to fly an aircraft. The design can incorporate this level of training and expertise in order to produce a product that is both safe to operate and that can be operated safely. A tool designed to be bought and used by a member of the general public should consider their general level of competence and awareness, and, in the absence of instigating a prescriptive training regime, the provision of full instructions with the tool might be considered suitable. Another important factor in considering this "safe to operate and operated safely" motif throughout the life of the product is how the administration (or management) of the product and its use is conceived.

Returning to the aircraft, pilots must undergo periodic re-training and assessment in a flight simulator to test their competence under all manner of operational conditions. This training is closely monitored against their actual flying time and is linked to their suitability for deployment on various aircraft. The planes themselves are subject to rigorous mechanical checks and servicing regimes that are, again, tightly controlled. The personnel undertaking such servicing are themselves highly trained and their work is closely monitored and cross-checked. It is not a case of coincidence that flying is the safest form of transport per passenger journey. Of course, any procedure or system can be subject to error, as the events surrounding BA flight 5390 in 1990 (AAIB, 1992) and the Costa Concordia cruise ship in 2012 (MCIB, 2013) are testament to. Human errors, or more specifically, errors of judgement, may not be entirely possible to design out, but it is possible to mitigate them, initially by understanding the range of outcomes that are possible through any given decision-making process. And, ultimately, through the proper reflection of events where loss or harm has occurred and instigating techniques in future products or systems to try to prevent a repeat of the same problem. Again, this is where use and design are mutually integrated.

The need for any design to incorporate specific training, use, or maintenance regime will stem from the originating statement of requirements, although perhaps not specifically. The client will create the statement of requirements with a view to fulfilling their ambitions, howsoever they may be derived. The client's ambitions may be to dominate in a particular product sector, to expand in a new market, or to create a product with class-leading sustainability, for example. The designer will then need to marry these ambitions, contained explicitly or implicitly in the statement of requirements, with any supplemental schemes such as training. Developing a new aircraft will clearly require specialist training for the operator, but a new model of motor car may need to ensure that the design fits within nationally or internationally accepted standards of practice. The latter is the reason that the symbols used on the switchgear in motor cars have become standardized over recent years. It is also worth noting that some manufacturers, like Porsche—who make exotic high-perfor-mance motor cars, offer their customers the opportunity to drive their chosen model on a racetrack in order to acclimatize themselves to the controls and functionality before using them on the public highway.

However a design accommodates the statement of requirements, it is important for it to understand all of the operational parameters (like those we discussed in the off-road lorry example) in order to ensure the safe use of the design's output throughout its life. Additionally, a design should consider abnormal modes of oper-ation as well as normal. Abnormal operation is where the product must function but with reduced safety systems in place; for example, whilst during maintenance or repair. The levels of training and experience that might be expected of a main-tainer or repairer will assist in designing appropriate and commensurate levels of safety during such abnormal use. In complex or safety-critical examples, such as a chemical plant or power station, the design may also need to consider exceptional operational circumstances. These are where all normal safety protocols have bro-ken down—for example, in the event of a fire or catastrophic failure—and where rescue parties, firefighters, or other highly trained personnel become involved with the product.

This may require the designer to put forward a number of conceptual solutions: for example, one that relies on artificial intelligence to monitor its operation; one that relies on training the operator and; one that relies on physical guarding of danger-ous elements of the design to protect the user. Subsequent to this, the designer can incorporate abnormal use—maintenance, repair, or running-in, for example—into the design in order to ensure the safety of operators other than those using it from day-to-day.

3.3 Implementing

Once our design has been finalized, and the product has been produced and put into service, the risks of its operation must be managed. Invariably, this is the remit of the safety adviser or safety department and will include operational parameters that fall under any of more than 100 safety-related pieces of legislation. Of course, if the design has been thoroughly substantiated against these pieces of legislation, or, at least, those that apply to the end product, then the safety function should be relatively straightforward. The fact that this is rarely the case suggests that perhaps not all designs are fully cognizant of the depth and breadth of safety legislation relat-ing to the use of all manner of things. An initial risk, therefore, in operating any new product is the realization that some form of adaptation may be required to enable the legal use of the product. Note the differentiation between *legal* use and *safe* use.

Adaptations may include such things as training, guarding, demarcation or the use of personal protective equipment.

Before anything can be used in the workplace, it must be subject to a risk assessment. This relatively straightforward task has perhaps lost some of its original meaning since the concept was finessed in health and safety law by the Management of Health and Safety at Work Regulations 1999, which requires a "suitable and sufficient assessment of … risks to … health and safety". The revised guidance to the regulations of 2013 refers now to the sexier-sounding "risk profiling" but the actual regulations which are still in force have not been changed (HSG65, 2013). Risk profiling is a slightly different discipline and the phrase is not as well known outside of professional risk management circles as the much more widely used "risk assessment". The object of the risk assessment is to identify any relevant hazards and consider the risks that might extend from them. A hazard of course is something with the potential to cause harm, and risk is the measure of that harm occurring in terms of likelihood and severity. For health and safety purposes, hazards are grouped into six categories, which are: safety hazards; biological hazards; chemical hazards; ergonomic hazards; physical hazards and; psychosocial (or organizational) hazards.

This assessment of the risks—which are now *observable* rather than *foreseeable*, as they were in the design phase—should provide a feedback loop for subsequent designs of a similar nature. Complex designs often cannot recognize every conceivable risk, and it should be beholden on those with a responsibility for safety to ensure that subsequent designs *do* incorporate relevant changes that improve either overall or specific safety concerns. Consider a number of injection moulding machines being designed and manufactured for a plastic manufacturing company. Upon installation of the first machine, it is noticed that, despite all the covers and guards that have been fitted to the machine, there is a space beneath the platens large enough for a person to stand up in. The platens are the large faces of the machine, like the jaws of a vice, that come together when the mould closes. As the machine is tall enough that someone could crawl beneath it, this exposes the risk that someone could potentially get into such a position as to be crushed by the platens as they close. If we imagine that this machine is capable of exerting 750 tonnes of clamping pressure—not an uncommon amount in respect of this sort of machine—it is easy to see why the safety adviser at the time might have due cause for concern. The simple remedy would be to fit a steel plate to the bottom of the machine to prevent anyone reaching up between the platens—a remedy that would have been far easier to achieve in the factory than at the client's location when the machine is fully installed.

This improvement to the safe use of the product, borne out of its actual use in the real world, as opposed to computer modelling or roundtable discussions, for example, could be considered the feedback loop of the Plan-Do-Check-Act cycle; although in this instance, it is probably more akin to the Plan-Do-*Study*-Act cycle that Dr Deming originally created (and preferred) (Deming Institute, 2020) (see Figure 3.3). It could

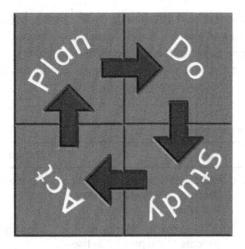

Figure 3.3 Deming's plan-do-study-act.

be part of the outcome of a "lessons learned" session at the end of the project or perhaps even part of a snagging review once the machine is installed. This demonstrates
that there are several ways this type of omission of safe use could be administratively
managed: whatever the project, industry, terminology, or layout of design and project
management, there is invariably some form of feedback loop that allows for this type
of crucial information to be fed back into subsequent designs. This feedback is demonstrated in Figure 3.4. Here we see that the expectation for the product, which drives
the original design, is tempered during the product's use to create new or alternative
approaches for subsequent designs.

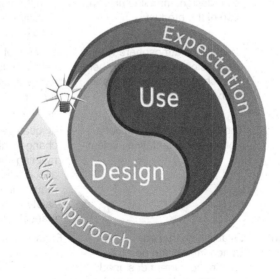

Figure 3.4 Expectation and feedback.

This feedback, howsoever it is produced, is, of course, highly dependent on the systems in place by the end user to establish, administrate and disseminate it. In our example above, we can see that there would have to have been a number of interventions taking place in order to bring about any amendments to future designs. We could imagine a range of these interventions in Table 3.1.

Table 3.1 Range of interventions

Discipline	Intervention	Output
Procurement	Verification that the product delivered precisely matches that which was ordered	Payment of suppliers' invoices
Engineering	Validation of the operational aspects of the product in line with the formalized expectations	Signing-off of the project pack
Health and safety	Inspection of the product for potential health and safety implications in its use	The creation of risk assessments on use, provision of work equipment, noise, manual handling, and so forth; and the creation of work instructions
Shop floor supervision	Full appreciation of the operational manual for the product	Assistance in the creation of work instructions
Senior management	The placement of the right individuals/groups to facilitate safe design, manufacture, and use of the product	The right people in the right place at the right time with the right information making the best decisions possible
Corporate culture	Creating an atmosphere where discussions can be freely held and challenges can be made without fear or favour	The prompt recognition of safety issues combined with their effective onward transmission to the appropriate authority
Document management	The creation and control of a robust documentation system	Adequate records of design changes, discussions, meeting minutes, specifications, statements of requirements, health and safety records, training manuals, and so forth
Design management	Establishing that any amendments are possible within the current designed operational limitations and function	Enhanced functional design

This list is, of course, not exhaustive and may vary widely depending on the organization involved. Large organizations may have more departments involved, and smaller organizations may have one person representing more than one discipline at a time. Neither of these situations is better than the other and, in fact, either can worsen the situation for different reasons. More people involved in any discussions can have a detrimental effect through there being a need for larger meetings, possibly of a longer duration, and possibly even achieving fewer results. The expression "too many cooks spoil the broth" still has resonance in today's fast-paced corporate world, and even when attempting to mitigate this by having one individual at meetings representing a number of stakeholders, it is still possible that some decisions might not fully represent the views of everyone. What is significant is the understanding of the cohesiveness of an organization in getting its product, whether this is a project, an item, an idea, or a service from the conceptual stage out to market. The primary functions of realizing this will always be influenced and controlled by secondary factors such as leadership and specialized support. This was described by Michael Porter in his model, which we have reimagined in Figure 3.5.

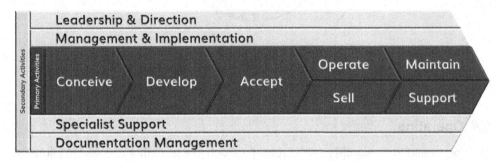

Figure 3.5 Adapted version of Porter's model.

In smaller organizations where one person may have several roles or functions to fulfil, it may be that their knowledge or skill set in any one of those roles is not as proficient as in others. The chief engineer who also has health and safety to look after, or the production manager who also acts as procurement head for example. In essence, the correct number of people engaged in the project or undertaking can only be determined by the project itself and, furthermore, a fundamental understanding of *what it is the project has to achieve in the circumstances*. This means it may not be wrong for the chief engineer to look after health and safety if the chief engineer *understands the implications* for health and safety in the project. The production manager *might* be best placed to run the procurement aspect of the project if it sits within their level of understanding and knowledge.

This is also pertinent to the continued use of the product throughout its functional life, and all this knowledge and administration (used in this sense to mean management) cumulatively forms the system by which not only safe operation can be achieved but also the appropriate procurement of future designs. An organization must understand the how, why, and where of its undertaking in order to ensure the procurement of the most effective solutions to its needs and ambitions. This may be thought of as the *organizational environment* in which the solution is designed to operate and encapsulates all of the disciplines of the organization that will come into contact with that solution, both to a greater or lesser extent. In Figure 3.6, we see some examples of how all these disciplines correlate.

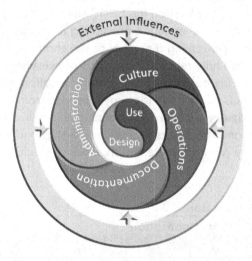

Figure 3.6 External and internal product influences.

3.4 Operating

The documentation system(s) must be robust enough to ensure that all relevant information is not only available but updated and disseminated appropriately. It is not enough to simply have copious amounts of information neatly filed away in project packs and hard drives. Informative documents should be (a) readily available to all those who need to review it; (b) regularly updated to reflect the latest position, and; (c) accessible by those who need to know that updates or amendments have been made. There are many complex and bespoke solutions to document handling but some systems can be equally well run from a simple spreadsheet. The correct approach is, once again, to understand the organization in which the system exists, who requires access to what information, and how that is best presented to them individually. A wonderfully exquisite and expensive software package of documents is unlikely to be of value to the overstretched shopfloor manager if they have no computer in their production hall.

The policies for documentation are often overlooked as "write and forget" documents which, once written, occupy a lonely and dusty corner of the filing system. They become "shelf-ware". This should not be the case. Policies are the core of the organization's systems in all departments. They set out the organization's stall in terms of ambitions and expectations. Policies reflect senior management's desires for the undertaking as a whole and, as a result, reflect their personalities and cultural aspirations. Polices are then used to determine sub-systems—such as processes and procedures—which, when followed correctly, will ensure these aspirations are met. Just as importantly, policies must be reviewed periodically to ensure that, firstly, they continue to reflect the organization's ambitions and, secondly, that they actually reflect the *current operational position* of the organization. This is important because any organization that is not adhering to its own policies is already in breach. This may not lead immediately to any serious repercussions but is certainly considerably harder to argue against at a later stage should anything go wrong.

The management of any organization is critical in ensuring that its ambitions are achieved within the timeframes and parameters set by senior management. Since perhaps the 70s and 80s, the discipline of management has seen innumerable incarnations and objectifications from a variety of probably well-meaning and linguistically-inventive sources. It matters not how an organization is managed so long as it is effective in the circumstances. The effectiveness of military management and planning should always be given respect in terms of its functionality, insofar as modern militaries have generally had centuries of practice at achieving desired outcomes in challenging environments. Perhaps only slightly exuberant business leaders might consider themselves equivalent to generals, but it remains that commercial organizations face similar types of issues in terms of wars, theatres of operations, battles, and individual objectives. All of these require different levels of resources, different types of skill sets, and different numbers of people to perform them. The various levels of documentation and communication can also be seen to be roughly equivalent too in terms of the policies, processes, procedures, and work instructions required at each level.

Planning how an organization deals internally with its management function—how it deals with the dissemination of information, the structure of communication, the processes, and the equipment of the work required—is also the basis upon which it can then deal with its operational aspects. Operations is where the work gets done—but without objectivity, guidelines, or structure, this would be wayward at best and impossible at worst to achieve. The maintenance of equipment only when someone remembers to do it risks either that equipment failing in operation, perhaps dangerously so, or being maintained too frequently at unnecessary cost. Only by firstly understanding the maintenance requirements in the design and then going on to formalize those requirements in a documentary system can effective and safe maintenance be managed. Similarly so, with abnormal operation as well as exceptions, such as emergencies and failures: the limitations of the originating design, the expectations of the organization, and the operational environment all play a part in formulating a robust and cohesive system for dealing with these types of circumstances. Abnormal and exceptional circumstances refer to any situation that is outside of the normal, day-to-day operation of any device or undertaking, such as repairs, maintenance, failure, fire, flood, collapse, and so forth.

All of these disciplines—documentation, management, and operations—are all connected to, and bound by, the culture of the organization. The culture is where such matters as how well and to what extent individuals are trained, for example, are envisaged. Other such functional matters as the methodology of procurement, competency of risk assessment, and methods of communication are as much a part of the culture as less-tangible things such as the encouragement of the welfare of the organization's workforce. There are perhaps readily imaginable real-world examples of organizations that utilize the latest thinking in robust management, with penny-perfect financial planning and profit maximization, and yet still suffer from excessive staff turnover due to poor welfare standards. Similarly, it can be the case that organizations can spend so much effort looking after people that they forget to make a profit—a situation that can only inevitably lead to organizational failure; although this is perhaps less likely than the contrasting previous example. Generally, people respond extremely well to being well-respected and cared for in any organization.

The culture of any organization is always a reflection of its senior leadership, for better or for worse. Senior leadership is ultimately responsible for assigning finances which can affect not just the quantity but also the quality of the resources that those finances procure. This is not just in relation to the physical equipment or machinery the organization might need but also the type, number, and competencies of the personnel it employs. This is also true of the *balance* of resources: having many more production workers than a given number of supervisors can adequately oversee is a case in point. It not only makes practical business sense to have an ideal operation/supervision balance but it is also prescribed in law in the Health and Safety at Work etc. Act 1974. This requires the adequate provision of plant, systems of work, instruction, information, training, and supervision to ensure the health and safety of employees. Clearly this reflects both the quantity *and* quality of those provisions.

Culture is also a predeterminate for the risk appetite of the organization. This can be reflected in such things as the investment decisions it makes as well as the quality of equipment that it procures or facilities that it occupies. Of course, buying inferior quality equipment is not always inextricably linked to safety issues, but it may well be the case that less-reliable or well-made equipment will need more maintenance or repair work. In itself, that is a financial decision for the organization's leadership and needs to be properly determined in a cost-benefit analysis, but it introduces the possibility of exposing personnel to more instances of abnormal or possibly exceptional operations. It is during these types of operations that there will invariably be an increased threat to safety simply because normal protocols are necessarily interrupted. Interestingly, it may not always be the case that procuring the very best or most expensive equipment will eradicate this issue either. Often, advanced, novel or exquisite equipment and machinery comes with a necessarily demanding or high-cost maintenance regime which may cause future issues should the organization be unable to sustain the financial burden. Not implementing a recommended maintenance regime can introduce safety issues just as readily as having to undertake maintenance too often.

The quantity of personnel and equipment can also affect safety, especially when an organization's appetite is to operate on a minimal provision of either, or both. The stresses induced on machinery when operated for prolonged periods at maximum production rates may be well understood by production engineers, but the same effects on personnel can often be missed. Today's better corporate understanding of employee welfare has undoubtedly improved matters but the corporate culture is still the defining paradigm for how that welfare is presented and maintained.

4 Human implications

4.1 Leadership

During the design and manufacture of any product, the group of individuals or organizations responsible for delivering it are (or at least should be!) well defined and controlled. Regular reviews throughout both phases allow the client to assess progress against clearly defined parameters. Exceptions can therefore be highlighted fairly rapidly before they become a crucial or even critical issue later on, and certainly before the product (whatever it may be) enters its in-use stage. In short, it is the defined parameters of the output, the expectations of it if you will, that guide and control the actuality, and in the consistent reviewing processes that take place, we are able to assess the adherence to, and continued relevance of, those guides and controls.

During the in-use stage, however, these guides and controls—as well as the ability to consistently assess them—may not always be possible. During this stage, we engage personnel to perform various duties connected with the product's use: this may be in the form of installers, operators, maintainers, repairers, and managers. The disciplines these types of personnel could belong to include engineering, administration, surveying, safety, and so forth. All of these disciplines and, indeed, often different role holders within each discipline, will have their own objectives, experiences, guidelines, and terminology which they will use to complete their duties. Each individual will also have their own personal objectives, experiences, and values to which their inner compass is rooted, and this will inevitably affect their own interaction with the rules, processes, and procedures with which they are expected to engage.

Above all this seemingly tangled web of roles, responsibilities, and machinations lies the senior management of the organization for whom all this in-use activity is taking place. Senior management will also be composed of individuals whose characters will be influenced by any number of traits, experiences, and biases similar to any of the other individuals throughout the organization. The overall constantly evolving soup of characteristics would, without proper control, descend into a homogenous mess were it not for formal ratification of the ground rules. Essentially, these are the policies, processes, and procedures with which we have become so familiar in the modern working environment. Without knowing precisely what it is we are supposed to be doing how do we know we are achieving it? And without knowing the standard that we must achieve in the work that we do, how do we know if we are doing it right?

DOI: 10.1201/9781003296928-4

These may appear to be rather facetious and obvious points to raise, but having an agreed plan is the only way to run any operation efficiently. And yet, time and again, we see situations where there is no agreed plan or where standards are not formally agreed at the outset. This will invariably lead to an unsafe event occurring. Just as important, however, is the ratification that the plan or standard *is being adhered to*. This, too, may seem blindingly obvious, and yet it is a situation that we find perhaps far more often than not having a plan in the first place. The reasons for this may be myriad but the core reason, we would suggest, is complacency. Complacency that a plan is in place, that documentation has been completed, that boxes have been ticked, and that all is well with the world. This type of complacency is rooted in the under-appreciation by senior management of the very basic psychology of the humans that they employ.

Let us imagine that in a factory, several pieces of steelwork must be welded together to provide the underpinnings of the next part of a manufacturing process. The steelwork will not be seen but is structurally important to the whole product when complete; therefore, it is not overly important what the weld looks like, but it is important that it is a good strong seam weld. A simple specification for the work can therefore be given to a welder to follow and if the welder deviates from the specification, the management would be within their rights to discipline them. The management might decide that replacing the welder with a robot is a more attractive solution to ensuring consistent quality. But what if the steel varies in quality? Due to the fact that it will not be seen in the final product, the management of the company may have decided to reduce costs by having the steel made to a lower standard of tolerances, in which case there may be variations in the fit during welding that requires the judgement of a skilled welder to accommodate. In this instance, it would be desirable to continue to use a human welder with their requisite skills, knowledge, and training and to allow them the autonomy to perform their work accordingly.

We see here that, in this instance, there are two solutions allowing two different levels of quality output from the same input. The input (the quality of steel) can of course be improved, in which case the consistency of robot welding will become more relevant. But this would have an additional cost burden which would make each piece more expensive: again, a decision of balances that the management must quantify and justify. The same variability of output can be seen in the handling of procedures in an organization, especially where safety procedures are concerned. Here, too, we can "robotize" the procedure by making them as encapsulating and detailed as possible, so that all we require individuals to do is follow them to the letter. This has been the system used by militaries for centuries: unwavering acceptance of the instructions that one is given to follow. But there are two important caveats; those being the quality of the input and the level of supervision. If the quality of the input—just like with the steel earlier—is not sufficient, or is of variable quality, in relation to the information contained in the procedure, then there is the danger that the output of the procedure will not be correct. We blindly follow the rules only to arrive at completely the wrong conclusion. This is where the level of supervision is crucial in providing a level of variability to the procedure, and in military terms this is where the corporals, sergeants, captains, and so forth become relevant; each with their own level of responsibility, authority, and training.

4.2 Trust and empowerment

The concept of small groups of people working cohesively and empowered with some level of autonomy in their job was realized by Elton Mayo during a number of workplace experiments conducted between 1924 and 1932. Although much of his findings have been subsequently challenged, the ideas around motivation, management cooperation and attitudes, and productivity being connected to employee welfare and participation, still resonate today. But the welfare of individuals in an organization needs to be managed in such a way as to be practical, thoughtful, and above all, meaningful. A bowl of fruit in the corridor may be the organization's response to "providing welfare" but is unlikely to produce the sort of results in employee participation and engagement expected if it is subject to derision by the employees.

Empowerment in fulfilling tasks can be an equally thorny issue if an organization feels that not controlling every aspect of their undertaking in fine detail will somehow impinge on the quality or cost of its output. This is where the controls surrounding the undertaking—the culture, management, expectations, and so forth—need to be firmly set in place. Like the markings of a tennis court, for example: how the players play the game is determined by their own judgements and skill, but there are rules about where they play and when the ball goes out of play. Consider the two stacks of stones, or cairns, in Figure 4.1 and imagine that the size of each stone is relative to

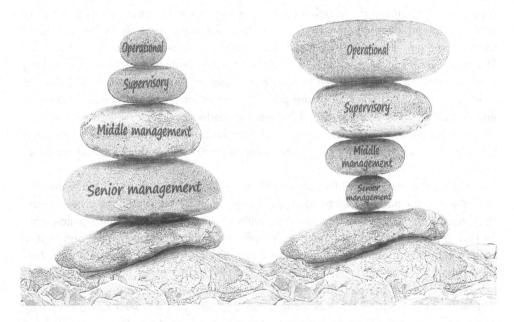

Figure 4.1 Comparison of the level of effort in organizational control.

the controls—of all types—in place for that function. The size is relative not only to the *volume* of controls but also the *level of thought*, or detail, that has been put into them. On the left is a representation of a robust, well-ordered organization that has in place strong cultural and managerial controls. These form the foundations for subsequently fewer and fewer controls leading to the operational part of the organization where they are so ingrained in the daily work function that individuals have the knowledge, and empowerment, to operate within them without an overly oppressive burden of direct control on them.

The cairn on the right is perhaps a more common and recognizable organizational setting, where the controls over individuals tend to be more relative to the number of individuals to whom they apply. Endless risk assessments, operations manuals, procedures, processes, work instructions, and so forth sit atop fewer and fewer well-defined policies, directives, and cultural ambitions. An abundance of operational dogma that can stifle the individual flair and creativity that might—just might—result in better productivity, lower costs, and improved safety. It should be evident how much more stable the one cairn is to the other.

The fear, however, of prosecution or litigation has led organizations to the type of model we see in the stones on the right. Implementing a safety system and then ensuring that everyone adheres to it is surely the central aim of the Health and Safety at Work etc. Act 1974? The thinking that we must have one plan to deal with safety and then many, many documents explaining this to all manner of individuals in the workforce seems to support this. And yet, time and again, when we conduct accident investigations, we regularly find that the primary cause is rooted firmly in the management of the organization. The risk assessments, method statements, work instructions, etc. that senior management thought were sufficient were, perhaps, not properly supported, implemented, or adhered to. The organization perhaps hired a safety professional and then left them to "get on with it" without ever really listening to them, supporting them, or bringing them into discussions with other elements of the business. The safety professional is left to encourage and implement safety without ever having any influence on what may well be the root causes of any potential safety failure. If management is cited so often as the primary reason for accidents, it makes sense to ensure that the most effort is expended here to understand how well controlled it is. Senior management has the most responsibility across the widest range of factors and influences on an organization, and this naturally will require the broadest range of control.

In Figure 4.2, we suggest the precept for beginning to bring about empowerment. Individuals will clearly be selected against criteria set by the ambitions of the organization's culture and trained accordingly within the remit of the training philosophy. The job or function should be specified, clearly, so that both the individual and organization know the ground rules (the tennis court may be laid out but is there deviation allowed on the type of rackets that can be used, the colour of the ball, or the weather conditions that the game can be played in). Individuals should be authorized by the organization to perform the task or function according to the level of training they have received, and this confirms the expectations made of them. These controls lead

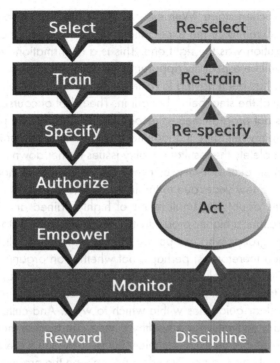

Figure 4.2 Precept for empowerment.

to the empowerment of the individual to fulfil their role effectively, although the organization will want to monitor that. The monitoring function acts as the feedback which provides the organization with information about whether their selection process is robust enough, that the training is suitably sufficient, or that the specification stage is quite clear. Exceptions to these verified expectations can then be dealt with in the traditional way: reward for those who excel or a disciplinary process for those who wilfully disengage with the system.

Rigorous selection, encouragement, and empowerment of people were identified in *Good to Great* as key drivers in organizations that had excelled (Collins, 2001). But the dichotomy of creating work procedures or even operating instructions is that the more exacting and precise they are (that is to say, the less autonomy they allow the follower of the procedure), the more they will rely on the quality of the inputs leading to them. This in turn requires the appropriate amount of control on those inputs which may, of course, involve other procedures and hence the inputs feeding into those. A method of controlling these inputs is obviously through the use of metrics: measurements to demonstrate that any particular product, procedure, or function is within the acceptable tolerances that have been determined for them. But metrics can bring their own issues depending on how they are initially set and the way in which they are recorded. Let us return to our example of the human welder versus the robot.

If the metric has been set by the production quality team, they may well be interested in the number of items completed by the robot welding machine and passed on to the next production stage. With the invariably high rate of output expected of robots, this may well reinforce the expectations of management that their investment in this high-tech solution was the right one. This is a confirmation bias on the part of management and will be especially apparent when compared to the output rate of the human welder who has been having to fettle several workpieces on each shift due to the variable quality of the steel being bought in. The robot of course has had no such issues because it is not programmed to recognize these in the first place. The metrics in this situation certainly reinforce management's decision to automate the process but possibly ignore completely the resultant quality issues further down the production line or when the customer takes delivery, or even when the product made from the steel pieces falls apart in a few years due to the failure of the weld seams.

There can be little doubt that small teams of highly trained and incentivized individuals often have either a higher productivity or higher-quality output, or both, when compared to larger groups who are poorly trained, tightly controlled, and have low morale. The question therefore is, perhaps, not whether an organization should consider empowering its individuals, but how it should go about it. Clearly it requires a set of fundamental factors to be in place, such as the right cultural aspirations and expectations, as well as clear guidelines within which to work. And culture and empowerment can be mutually self-fulfilling: a healthy culture can lead to more empowerment which in itself can promote an even healthier culture. This in turn creates better worker welfare which can only have a positive effect not only on the organization but also on the wider society in which it operates.

4.3 Safety in numbers?

Safety metrics are equally subject to this conundrum of misinformation in that their verisimilitude is based entirely on how they are set and measured. Additionally, how they are presented to senior management and, equally, how senior management review them can also lead to a self-fulfilling prophecy of satisfaction that belies (or even completely ignores) the true situation.

Consider a manufacturing company that has called in a specialist contractor to assess and advise on legionella. The contractor attends the factory and conducts a full assessment. The final report notes just two issues: a chemical dosing system connected to the mains water that has no non-return valve (NRV) and a dead-leg on a hot water pipe where a wash basin has been removed in the past. The contractor recommends, correctly, that an NRV is fitted to the chemical dosing manifold and that the dead-leg tee connector is removed and replaced with an elbow joint. The report goes to the health and safety department who sends each recommendation to the two separate facilities managers who look after each location where the issues are. Once done, the health and safety department note that their job is complete—the legionella assessment is complete, on time, and within budget, and the report has been returned—again, on time—and the recommendations have been disseminated to the correct location managers.

The issue is that neither facility manager has any idea what the terms "non-return valve" or "dead-leg" mean. They were not present when the assessment was carried out, and although they have heard of legionnaires disease, they don't really know how this is connected to what they do in their particular part of the factory. Added to which, the report from the contractor is filed by the health and safety department because, as far as anyone is concerned, the information has been passed on and is, therefore, "dealt with". The facility managers do not get around to contacting the maintenance department to get the work booked in because they file the paperwork under their "To Do" pile, and therefore it never gets looked at again. Meanwhile, in a year's time, the contractor returns to conduct their annual inspection again, notes the need for an NRV and removal of a dead-leg, generates a report, and the cycle continues. All the while, of course, the metrics for reporting and conducting annual assessments are passed with flying colours.

This may seem a fairly trite illustration, but it is a real-world example that the authors have experienced at a client's location. The cycle was broken by the authors asking impertinent questions like "why hasn't this work been completed?" and watching the resultant scurrying for answers. There is a saying, promoted even by the HSE, that "what gets measured gets done", but this is not always strictly true. Certainly, it is important to measure an organization's performance, especially in safety, to ensure that we can see firstly where resources need to be focussed and secondly where progress is being made. The issue is that in some organizational cultures, it is the pursuit of good news that skews the results. If senior management is unlikely to take kindly to hearing bad news, then the person tasked with delivering it is likely to take a view on how that news is communicated; or indeed how much of it is communicated. Senior management should always be prepared for bad news and be professional enough to deal with it appropriately. Shooting the messenger should not be a recognized organizational practice; certainly not when trying to develop trust within an organization.

4.4 Competence

All of this ties in with another factor over which senior management has absolute control—the level of competence within the organization. It is a truth (that should be) universally acknowledged that an organization in possession of a poor culture is in want of a competent manager, to paraphrase Jane Austen. Culture is at the heart of everything an organization does and the staff within it will feed off that culture, irrespective of their own individual characters and experiences. A culture that is ignorant of, or indolent about, safety is as bad as one that is completely consumed by it. Both will create an atmosphere of fear, particularly among those responsible for safety for either saying too much or not enough. As with all things, there must be balance. And there must be competency in driving competency. Senior management must ensure that the right people are put in the right places for the right reasons. This can often be difficult in smaller organizations where some individuals are responsible for several disciplines simultaneously, and, as organizations grow organically, this can lead to legacy issues where individuals are in a particular position due to their length of service or some other historical reason.

Training is, of course, key to competency, but it should not be forgotten that the *overall* measure of competency is abbreviated as SKATE: skills, knowledge, aptitude, training, and experience. The aptitude measure here is vital when looking at legacy issues in organizations because training or re-training staff in order to fulfil newly created or newly focused roles in a growing business is only salient if the individual in question has the requisite aptitude for the role. Training to be a doctor is, like any training, only about the accumulation of knowledge, but prerequisites would certainly be empathetic skills with patients and a strong stomach.

One expression that usually raises the eyebrow of the seasoned safety professional when querying or assessing a work operation is "we've always done it this way". Not simply because of the dismissiveness of the statement but because our knowledge of the mechanics of safety suggests that the absence of an adverse event does not necessarily demonstrate inherent safety. The oft-quoted "Swiss cheese" model attributed to James Reason demonstrates this all too clearly—the holes in the slices of cheese, representing flawed control measures, must align in order to let the threat of risk through. This may take moments or it may take years, depending on the undertaking and the inherent robustness of the controls and the competency of the individuals performing the task. The point is, failure in this context is *probable* at some stage in the future. Saying "we've always done it this way" also suggests a lack of willingness to engage with the process of discussing or examining better ways of working, although it should be borne in mind that it is always possible that the best methodology may have already been arrived at. The "way that things have always been done" can also inculcate a resistance to ensuring that the output of a department, or even a whole organization, is the best it can be—in terms of safety and quality—for either the next stage of processing or the end user. This is often the result of "silo thinking"—where each department in the chain of production is liable to resolve issues within their own universe of thought and experience, without subsuming those of other departments involved in the overall process, or even the end user.

Another way of saying "we've always done it this way" is "well, we know what we're doing". This is a common design error that can lead to potentially harmful outcomes. The design of complex systems for use by technically trained operatives is acceptable if the training includes a suitable explanation of the technical complexities. But those same systems designed for everyday operators must be cognizant of the potential lack of technical skill. Computer operating software is a case in point where, for reasons of global domination, it is aimed very much at the level of operator who just wishes to switch on their computer and work. The problem arises when trying to fix problems that occur, and the solution involves lengthy and technical explanations from the manufacturer, using programmer jargon that for the vast majority of computer users is beyond comprehension. The outcome of this may be the loss of important data, but the consequences of silo thinking can be far worse.

Many large organizations rely on sometimes fairly complex process chains to get from the point of a customer ordering a product to that product actually arriving at the customer's location. This may be of little surprise, but sometimes where organizations have grown organically there are components within the chain that have become overlooked, often for reasons of stability, complacency, or sheer lack of interest. It can be difficult to enthuse about the mundanity of the packing process when conjuring up an award for the latest hot-shot salesman for bringing in that colossal order that is going to save the business. But each component in the production chain is as important as any other—the system is cohesive and interdependent. Many businesses fail because they take on orders they cannot fulfil; or open in new markets without having the necessary supply chain to bring stock to it. Similarly, armies fail because their advancing troops have advanced quicker than their supply lines; or individual processes fail because a component or process either side of it is not fit for purpose.

Case study—Kegworth Air Disaster (1984)

A Boeing 737 was approaching East Midlands Airport near Kegworth, Leicestershire, having been cleared for an emergency landing after suffering an engine failure shortly after take-off. Remarkably, the pilot subsequently shut down the wrong engine, i.e., the one that was still working, due primarily to poor instrumentation in the cockpit that did not clearly identify one engine from the other. This newly designed instrumentation layout had been introduced in this upgraded version of the aircraft without sufficient training, including the use of simulation training. In addition, there was no standardized approach in aviation to identifying engine positions: the left-side engine was variously known as left, No. 1, or port side engine; the right-sided one as right, No. 2, or starboard engine. This *belief* that everyone else in the chain of production and use would simply understand what was being displayed and how it was referred to led to the deaths of thirty-nine people and a further eight from their injuries (AAIB, 1990). More remarkable is that, years later, Boeing would again introduce systems, ostensibly to promote safer flight, which, due to the lack of engagement with, or explanation to, the flight crew, would lead to another aircraft loss, this time of a 737 Max.

This is where organizations might develop a contingency plan in order to provide some resilience should some disastrous event take place. Business continuity is an important aspect for any organization to consider in order to demonstrate preparedness, but such planning is almost always concerned with some *external* force or influence: power outage, fire, flood, environmental changes, and so forth. A risk register will have identified these types of risks to the organization, and measures to control or mitigate the consequences will stem from this. What the risk register is less likely to have identified is exposure to risks *inherent* in the organization, such as human influences, poor decision-making, or loss of staff morale or loyalty. We could suggest there are two reasons for this. Firstly, the organization does not even know what risks it is exposed to within itself, and secondly, no one wants to think that the root cause of failure might be right underneath their nose.

A different philosophy for dealing with resilience is that of "antifragility", proposed by Nassim Nicholas Taleb (Taleb, 2012). Instead of developing a system of resilience that deflects the consequences of risks to which we are exposed, an antifragile system absorbs the consequences and becomes stronger for it. It could be compared in some way to the common expressions "learning from your mistakes" or "what doesn't kill you makes you stronger". The problem with an antifragile system is that it must have failed at some point in order to have become more resilient, which raises questions when considered in the context of safety. An example might be a project manager who has had several large projects fail may be more likely to be better at preventing failure in the future than a newly qualified project manager. There may be some truth in this; indeed, one must have failed at something in order to understand where the pitfalls are in the future.

The issue here is that we have in the first project manager someone who has failed before and therefore has a poor track record, notwithstanding the fact that the failed projects would undoubtedly have led to losses, which would not have been terribly agreeable for those involved with them. Biological systems have been shown to demonstrate antifragile tendencies, and this also leads us to the worrying prospect that, perhaps, the correct action of governments to the pandemic of 2021/22 would have been to have let the world's population just get on with the business of becoming infected and developing immunity naturally. This is known in epidemiological circles as "herd immunity". The human immune system is remarkably good at developing resistance to new forms of infection, but only over time. This can have devastating consequences for the population at large but results in better immunity overall, such as during the outbreak of bubonic fever or "Black Death" in the mid-14th century. An understandable lack of government response may have been inevitable in the Middle Ages but is highly unlikely to have garnered much support today. Irrespective of its merits, we would submit that even an antifragile system of risk management still requires a deep understanding of *where the risks are*, both internally and externally.

In consideration of growth, we should look at any mammal at birth. It is generally in proportion; a miniature version of the animal it will become. Again, as it grows, it does so generally in proportion. Were the arms or front legs to grow disproportionately quickly, for example, then the chest muscles would not be able to support them, and the animal would not be able to walk, run, or feed. Business growth should ideally follow this pattern, albeit with some flexibility in each component. The flexibility allows for bigger and bigger orders to be taken as the rest of the production chain catches up. Similarly, having flexibility allows for later stages of the chain to be reduced in output if orders dry up for a period of time. Steel mills are an example of where this type of flexibility is not possible as their furnaces can take weeks to both reach the required temperature and cool down again sufficiently following a shutdown.

Quality standards programmes are an example of where great efforts are made to achieve high levels of product quality at the production stage and yet can fail to examine in sufficient detail other components in the chain. The sales team may provide the most amazing service in delivering on the customer's needs, but does the customer service team deliver an equivalent level of service should something not be quite right with the order? Promising the world and failing to deliver is something that only a few of the very largest global corporations can consistently get away with. Packing, again, is an area that can be neglected—how many companies for instance mail their own product to themselves in order to make sure that it arrives safe and sound?

An organization's inability to foresee potential issues with any and all of the components within its production chain can lead to quite disastrous outcomes stemming from often fairly benign issues. How many organizations rely on the knowledge of just one person to operate a mission-critical piece of machinery that forms part of the production chain? Should that person become ill, or leave, or retire, who will operate the machine in their stead? And to what degree of competency? But this is not just an organizational issue—entire nations can suffer from the same issues. The lack of skills training in the UK throughout the late 20th and early 21st century has led to a crisis of competent workers in many sectors—such as engineering, nursing, teaching, and logistics—which was exacerbated following the coronavirus pandemic and the UK's earlier leaving of the European Union. This risk blindness is generally only acknowledged with hindsight after the event—but hindsight is not a risk management tool.

Case Study—Bhopal (1984)

Approximately 40 tons of methyl isocyanate (MIC) gas, mixed with other unknown gases, leaked from a chemical plant owned and operated by the Indian division of Union Carbide causing some two thousand fatalities and a further estimated two-hundred to three-hundred serious injuries (Rosencranz, 1988).

The Union Carbide plant in Bhopal was failing due to a problematic process of producing fertilizer from an alpha-naphthol process, and by 1984 the plant was at just 20% capacity. Remaining stocks of chemicals were combined to make approximately 62 tons of methyl isocyanate, 22 tons of which were pumped into tank 611, and the remaining 40 tons were pumped into tank 610. Production should have originated from tank 610, but due to pressurizing issues, tank 611 was used instead.

On 2nd December, the plant pipework was washed down to reduce the effects of corrosion of the methyl isocyanate. Prior to washing down, a blanking plate was meant to be fitted in a coupling to prevent the wash-down water from entering areas where it might pose an issue. This task normally took a couple of hours to achieve, and it was not unusual to have trace chemicals fall out of the open pipe onto the operator during this time. It became normal practice, therefore, not to fit these blanking plates due to the required level of effort (Labib & Champaneri, 2012).

Most of the safety systems were in a poor state of repair, including the vent scrubber which was there to neutralize any gas leaks, and the flare stack for flaming-off gases. The fridge plant designed to keep the methyl isocyanate cool had been shut down so that the refrigerant gas could be used elsewhere, and the fire water spray system could not cool the flare stack since it could not reach the top. The vent scrubber, designed to filter the gases through caustic soda rendering them inert, was also out of action due to maintenance (García-Serna, Martínez, & Cocero, 2007).

Despite the poor state of the plant and supervisory personnel ignoring several safety features—such as the fitment of the blanking plate—the operations proceeded. Wash-down water entered tank 610 via a valve that had either not been shut or was failing to seal correctly, and this mixed with the methyl isocyanate in the tank causing an exothermic reaction which increased the internal pressure from 2 to 10 psi within minutes. This rise in pressure was ignored as the operators were used to spurious readings from faulty gauges. The pressure in tank 610 soon reached 45 psi, well beyond its operational limits. The blow-out disk (designed to fail at 40 psi) ruptured, allowing gas to travel along the pipework to the vent scrubber. The control room operator went out to check the process and, upon hearing gas escaping, ordered the control room to engage the vent scrubber with the caustic soda, but this failed to happen due to being out of service for maintenance. The alarm was sounded and then silenced moments later and poisonous gas continued to escape for two hours (Browning, 1993).

Additionally, the pandemic brought into sharp relief the woeful lack of preparation that many nations had for a cohesive response to it. This, despite the fact that pandemics feature highly (or at least should) on any state's risk register. However, foresight is one thing; proper preparation is something quite different. This is where not only accurately identifying the likelihood and severity of a risk are crucial, but also the *velocity* of the severity: the rate at which the threat escalates. Simply acknowledging a risk by drawing it up in a risk register is not sufficient; it must be properly prepared for or otherwise it is just another form of risk blindness. Similar to accepting that just because Bob in packing has been operating his machine for 20 years, he is going to be doing so for another 20 years without question.

Preparing properly for the risks faced by any organization is a culmination of at least three factors:

- The proper identification of salient risks.
- The input of the correct information.
- The right people assessing all this data and making informed decisions.

The growth in any organization, from its nascent beginnings to world domination, is rarely a stratified process but more generally an organic one, where multiple-role holders become more specialized as more and more personnel are brought in to administer busier and busier departments. These individuals bring with them their own experiences and expertise and eventually the original culture of the business begins to dilute on a micro scale. On the surface—and certainly in terms of the public perception of the business—the organization may well appear to be firmly adhering to its founder's guiding principles but, internally, inevitable power struggles and politics act as stressors on the culture. As the individual departments in an organization grow and adapt, they can become ever more dislocated from one another, with each following guidelines, principles, and standards applicable to their particular discipline.

During early growth, as different disciplines in a business are managed often by perhaps only one or two individuals, they tend to "get by" as best they can. The founder of a business who is, perhaps, an extraordinary salesman, may approach doing the accounts with a fairly blasé attitude until such time as the accounts become complex enough to warrant having a full-time administrator. Over time, as the business grows further, there will come a time when another accounts administrator will be needed and, at some further point in the future, these two individuals will need to have a manager. The accounts staff become a department and start to have their own identity. In time, the business may grow to a point where it employs its own qualified accountant or financial director. Accounts is now a fully fledged function of the organization with its own terminology, documentation, management structure, and office parties.

Having dedicated expertise in accounts means that the organization can be better informed of relevant financial regulations, opportunities, and threats. Financial risks can be better understood and evaluated, and the management of those risks can be more adroitly performed. Where the founder of the business once did the accounts on a Sunday afternoon at the dining room table with a sense of resigned acceptance that they had to be done, the organization now has a robust, dynamic financial team that can guide, steer, and cajole other departments quite effectively. Everyone, remember, follows the money.

This slow process of dislocation of the accounts department within an organization will be happening with other departments too. Sales, marketing, research and development, dispatch, safety—every discipline may start to take on its own identity and, indeed, culture. Conversely, the measured organic growth of an organization can also be combined with individuals being promoted—either by design or by stealth—simply because they have been there from the beginning, so to speak. This can also lead to cultural anomalies as these types of individuals can sometimes hold on to the early concepts and motives of the organization without changing and bending towards new practices and ambitions. This can lead to some departments within an organization appearing to have quite different objectives and management styles to others—perhaps in relation to those that have been developed over time during periods of expansion.

4.5 Changes over time

The Greiner model of business growth suggests there are several transformation points that an organization goes through during its life as it grows and evolves (see Figure 4.3) (Greiner, 1997). At each of these points, there is a change required in the paradigm which will allow the organization to flourish in the next phase, and so on to ever greater growth. Failure to adapt to these requirements, it is suggested, changes these transformation points to crisis points which can harm the organization's chances of continued growth and development. It is at or during these transformational points that

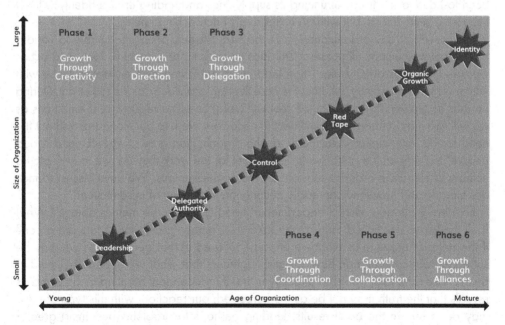

Figure 4.3 Greiner's model of organizational growth.

deviations away from the original ambitions and accepted values of an organization can occur, either by design or through unchecked evolution. That is to say, whether the individuals responsible for the change are appointed as part of a concerted effort to change—perhaps due to an investment into the growth of the business—or whether it is due to the elevation of some individuals in the management chain simply due to their length of service. The former tends to be more readily identified due to a possibly pronounced deviation from the existing culture by the new, incoming head of department: "nothing sweeps like a new broom" is the saying that readily describes this. The latter type, that of more pervasive deviation from the organization's cultural ambition, tends to be less easy to identify because the cultures of the new and the old diverge more slowly. The organization can develop ambitions that begin to leave the "old guard" behind: their resilience to change becoming ever more prominent.

Of course, in a rapidly expanding organization, it is possible that each phase presented in Greiner's model has to be tackled almost consecutively; each one appearing before the dust has settled on the last. This can cause "holes" to develop in the organizational framework—holes that tend to be quickly filled with temporary solutions that can, over time, become permanent ones. The "runaway train" syndrome might sound appealing from a financial aspect—it is certainly unlikely to cause despondency among shareholders and those due a bonus—but it is likely to cause significant failures in the communicative effectiveness of the organization. It may even cause a fundamental shift from the original ambitions and cultural stance of the organization, something that we have seen in several fast-growing, and now vast, internet-based companies. In essence, this is the same as identifying the velocity vector of any other risk in business and is something that military thinking again already understands. Many battles in history have been lost due to an army outrunning its supply lines and finding itself suddenly without support or ammunition, but with an increasingly encroaching enemy.

The growth of any organization is, in the early stages, often attributable to various biases that we humans all possess. Our bias towards those who think, look, and act as we do—known as affinity bias—can lead us to employing and promoting those with whom we have an affinity, whether they are the right *individual* for the job or not. During periods of growth, those whom we feel will have the same regards and ambitions as we will, we may surmise, will continue the necessary work to our standards without the need for our intervention. Obviously, as the organization grows, we will need "good people" to carry on the good work. To account for the work they do, we may deploy a number of metrics in order to substantiate their achievements. We have discussed metrics already and how they can easily become an instrument of misdirection.

So what if our new, dutiful departmental head is setting the metrics ablaze? What if all the boxes are ticked, and all the RAG (red/amber/green) indicators are green? If the statistics prove to us what we already believed in that person, then we may be suffering from confirmation bias: the prejudgement that what we already believed to be the case is, in fact, true. The difficulty is that if independent auditing of the numbers reveals that the truth may not be as we perceived, our reaction, with this type of bias, may be to ignore the audit results, leading perhaps to problems becoming greater or more ingrained. Or we may suffer from another type of bias: that of attribution.

Attribution bias is where we tend to attribute successes to those with whom we share an affinity and, conversely, we attribute failures to external causes. An independent auditor to whom we owe no allegiance is, therefore, likely to be at fault in this example: bringing us clearly bad news about a loyal employee or department in whom we have vested so much trust and faith. Not to mention time and expense. The difficulty facing senior management is appearing to have made the wrong decision: to have "backed the wrong horse", as it were. But to err is human, and if an organization is to flourish properly, appropriately, and—above all—safely, it must understand what errors have occurred, where they have occurred, and then correct them. A statutory duty under the Companies Act 2006 for directors is to "promote the success of the company for the benefit of its members" (Companies Act, 2006): improving an organization through the timely identification and mitigation of faults, however embarrassing it may be, certainly counts as that.

Undoubtedly, during phases of growth is where an organization can most readily find itself at risk of losing sight of its safety culture, whether intentionally or not. The crisis points detailed in Greiner's model are certainly high-risk areas for maintaining this stable safety culture, but they are not exclusively so. Safety—and perhaps more so, health and safety—has become far more entrenched in boardroom discussions over the years, but with it there has come the risk that it can develop a level of complacency. We do not believe this complacency is due to some dereliction of the duty of care that senior leaders owe to their organization and the people it employs: as we have seen, it is possible that complacency creeps into an organization slowly and unwittingly. If directors have attended their director's safety awareness courses, have appointed those whom they consider to be appropriate individuals to safety-related positions, and have implemented systems of reporting and metrics, then what more can they do? If everything is being reported as "green" to them, then surely all is well, isn't it? The important thing here is to "see beyond the greens," which means checking the validity of the information being reported to board or senior management level; something that requires far more than simply asking, "are you sure?"

The development of an organization over time is not only influenced by periodic stress fractures. There are many influences that will have an effect on the organization as a whole, or even just parts of it, depending on its size, spread, and function. In this respect, it is not wholly different to the influences that affect the evolutionary growth, or natural selection, of animal species. If we take humans as an example, we note the internal, autonomic evolution of defences against infections and diseases, as well as the hard-learnt skills of creating tools, transport, and structures to protect—and stay protected from—predators and environmental conditions. Some factors could not be foreseen for a long time, such as the effects of bacterial infection for example; whereas others we have known about for much of human history—fire, for example, or living close to tidal water. Organizationally, we also have influencing factors that we may not be able to predict but could certainly be aware of, such as the ever-increasing threat of environmental impacts. Other internal influences should be much easier to foresee and account for. Figure 4.4 is an adaptation of a diagram demonstrating evolutionary selection factors which compares the evolutionary impacts to humans

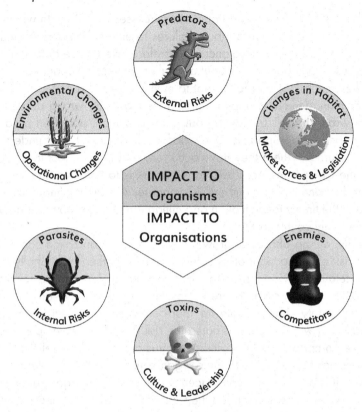

Figure 4.4 Evolutionary vs. organizational impacts.

with those of potential growth impacts to an organization (Kurzgesagt, 2013). These would make excellent starting points for listing what factors *might* affect one's own organization in order to begin to understand their possible impactive effects.

The culture of an organization should be one where senior management engages in periodic shopfloor visits or tours. Not simply to check all is well (and fully expecting the answer that it is); nor to cajole or berate those whom it is felt are not working as they should. This is not the Victorian era, and senior managers who behave like dominating mill owners are unlikely to be nurturing an appropriate safety culture or even creating one. Witnessing firsthand how the daily tasks of an organization are marshalled, supervised, and completed can give a holistic view of the entire undertaking that is often not possible by just reading reams of reports and metrics. It can also provide an abundance of relevant questions that might be asked of those managing that undertaking; questions that can give depth and meaning to the reports being proffered by them.

4.6 Financial accountabilities

As organizations grow, so too do the systems and processes that control their function. This is entirely normal and to be expected. Let us take the example of budgets. In a very small organization, there may be one manager who oversees pretty much everything on behalf of the company owners who are out and about generating new business for a lot of the time. If the manager needs to buy some equipment, or invest in a new piece of machinery or a delivery vehicle, then it is the work of a moment to discuss this with the owners and agree a way forward. The manager might have a company credit card with which to purchase small sundry items like protective clothing or packaging supplies. Whatever the manager spends is almost immediately observed by the owners and therefore the manager must be able to readily account for it. Ultimately, the owners have placed their trust in the manager to operate the organization who is therefore entrusted with a great deal more than the value of a few small purchases.

In a very large organization, there may be dozens of managers each with their own departments, responsibilities, and requirements. There are most likely to be various seniorities of management too. Their responsibilities will be varied, as too will their length of service and accountability to those above them. Possibly only the most senior managers ever get to sit at the same table as the owners or directors and, almost certainly, casual chats about requisite purchases will be few and far between. This obviously calls for a more structured approach and, therefore, it is likely that each manager will have a budget within which to make purchases for their relative department within strict guidelines. The budgets will be set by the financial director along the lines of the departments' needs and previous spending habits.

The issue can be, of course, that budgets are there to be spent, and if they are not spent, then they tend to get reduced in the following year's spending review. It is not wholly remarkable that many organizations, both private and public, have a spending spree in February and March as they desperately use up their allotted budgets before the financial year-end in order to make sure they do not have their budgets cut the following year. This system of financial control is, at one and the same time, both efficacious and highly inefficient. And it is not confined to finances.

4.7 Safety systems and safety accountabilities

Safety systems have, over time, been developed to reduce the reliance on shop floor level decision-making. Instead, they have promoted the concept of simple process flows; a sort of "connect the dots" of safety, or what some might ungraciously call a "tick-box exercise". The reasoning behind this type of system is clear: it allows a greater number of people to provide safety feedback with the least amount of training required. You may recall the maxim "what gets measured gets done" (Henderson, 2015), a quote variously attributed to Peter Drucker, Tom DeMarco, Lord Kelvin, and even the Renaissance astronomer Rhäticus (Cornell, 2007). The origins of the saying are, in essence, that if one is able to measure something, then one can manage it. It is a staple of the health and safety profession, being as it was quoted in the now supplanted version of HSG65 *Successful health and safety management* (HSG65, 1997).

This previous version of HSG65 promoted a methodology for a safety management system abbreviated as POPIMAR, which stood for Policy, Organization, Planning and Implementation, Measuring, Auditing and Review, although auditing was arranged as a feedback loop into the preceding sections. The mnemonic was a useful way for people to focus on the important factors that needed to be considered in the composition of a robust safety system, starting as it did with the policy, which, of course, would influence everything else that came afterwards. But it was felt that it was too prescriptive; too "health and safety" oriented. In wider business and operational terms, and for considerably longer, the methodology of Plan-Do-Check-Act (PDCA), created by Dr Deming in 1943, was more widely understood. PDCA was of course a corruption of Deming's original Plan-Do-*Study*-Act, which he felt was more impactive than merely *checking* something. The *Measuring* in POPIMAR provided the proper impetus to ensure that a safety system was performing as intended in a way that perhaps just studying or checking does not infer. But PDCA was understood by far more professions than just health and safety, and thus it was intended that advice on safety management was better aligned with the management of other aspects of an organization's undertaking. It is to be applauded that safety should be integrated into the generality of business management—safety after all should not be considered as some sort of "add on" to how everything else in an organization is conducted. It may have had a detrimental effect, however, on some people not implementing an adequate *structure* for a management system, let alone one specifically for safety.

The measurements of a safety system are best made at its "sharp end": that is to say, where the work is done. And providing those who supervise and manage that work with the proper tools to make those measurements makes a great deal of sense. Unfortunately, just like our budget example earlier, this can be both a blessing and a curse. Reducing the safety message—the very practice itself of ensuring the safety of individuals and the work process—to a line-by-line check sheet creates a number of issues, such as:

- The individual performing the checks is not empowered to understand what it is they are doing, or why, because of a lack of training.
- The organization does not get valuable information due to the lack of practical and specific training on identifying possible risk.
- The system can be developed in any number of ways to skew the information, either positively or negatively.
- Engagement with the core reasoning of the system by the workforce is subdued because individuals only see an impractical check sheet rather than the overriding ambition of the organization towards their safety.
- The recording of any exceptions to any list on a check sheet—any situation or risk that is not specifically covered—is entirely dependent on the whim of the individual carrying out the checks. A task for which they are, again, probably not trained.

In essence, the "check list" forms part of a deontological system, whereby the individual completing it is duty-bound to select an item (or tick a box) because it is the right thing to do in terms of their training and their responsibility to the organization for whom they work. The system itself, however, is ethically consequentialist in that the methodology chosen—in this case, the type and range of responses possible—are determined by what the outcome should look like. We can shape the outcome by ensuring the inputs are of a type and style we are most familiar with. This is not necessarily malicious, as we shall discuss later, but merely the result of the human behaviour of confirmation bias.

The gradual erosion, for want of a better word, of the safety system has been exacerbated over the years by several factors, not least the one we have already mentioned: the replacement of POPIMAR with PDCA. POPIMAR was learnt by safety professionals and used by them to effectively arrange and manage those systems. Its replacement with PDCA in 2013 was in part due to a raft of changes and perceptions primarily in health and safety at the time from government and academia in the UK. The argument that PDCA was far more widely understood and that the process of safety management could be more readily dovetailed into other organizational functions, perhaps engendered the hope that health and safety was not a "dark art", only to be practised by the initiated, and that it would create a more inclusive environment in which safety could flourish. The downside, however, is that by simplifying it, there was a danger that the core messages and ambitions of safety management would be simplified too. And possibly misunderstood.

Further qualification of the PDCA approach came from ISO standards such as 9001, which began to be standardized around a framework of principles, including PDCA. It could be suggested that as compliance managers began to see that safety systems were attributing the same principles as those that they understood in areas such as quality then they would begin to see safety within their remit. The subsequent conversion of the health and safety standard BS OHSAS 18001 to ISO 45001 clearly added to this perception. This is not to imply that an organization following an accredited standard for any of its functional systems is a bad thing. But it should be remembered that ISO standards *in themselves* do not regulate the quality of the information being recorded, merely the *way* that information is recorded and disseminated. It still takes an experienced eye to identify a hazard, understand the risks that emanate from it and, most importantly, implement effective control measures *relevant to the environment in which that hazard exists*.

We should note at this stage that in line with Dr Deming we concur that plan, do, *study*, act has a much more inquisitive tone than plan, do, *check*, act. Certainly in line with the observations we shall be making later the point of studying the information we have garnered is more important—and far more likely of delivering better informational results—than simply checking that something has been achieved or noted. It falls once more into the realm of phraseology and accepted practice, but we must counter the notion that we are performing a simple tick-box exercise when, in actuality, we should be examining the detail of the "planning" and "doing" we have recorded. This will become more evident later in the book.

4.8 Reliance on software

Additionally, in recent times we have seen a profusion of software programs designed solely around safety systems. These programs are designed to help with such things as the assessment of risk, the auditing of processes, the management of safety-related functions and the compiling of construction phase plans. Again, there is nothing essentially wrong with digitizing the process of safety, but we should be wary about what information is recorded and the way in which it is disseminated. Having more people "at the coal face", as it were, observing and recording safety-related data can only be a positive thing, but there is the risk that the *act* of compiling this data becomes more significant than the *reason* for it in the first place. As with any measuring or recording methodology, the output from it must be viewed with a critical—and above all, trained—eye. We must resist the notion that just because we have measured something our duty is done. Instead, we must thoroughly understand those measurements and act on them accordingly. As with the issue of metrics that we have already discussed, the setting of parameters within a software program can be all too easily influenced by our own (or the software designer's) standards, experiences, and expectations. If these conveniently meet the inputs that are fed into the software then we are in danger of being misled by its output.

4.9 Human factors and ergonomics

Having a standard to work to, which is then audited and accredited by a professional body, is an important factor for organizations, particularly as they become larger and more operationally fractured over time. Ensuring there is a "single vision" for everyone to see and work towards can lead to an enveloping sense of culture and ambition. But compliance is exactly that: just compliance. Whereas the most consequential things about the information—or measurements—that we have discussed as being so critical to ensuring the continued safety of an organization, function, or asset are the *quality* and *relevance* of that information. It is here that the bureaucracy of compliance can show its hand, as with any system where human implications are involved. Consider an alarm system in a chemical complex which displays alarm notification as it is encountered as a line of text on a monitor screen. Each successive alarm pops up at the top of the screen, subsequently pushing the last top-line alarm down one line. As time goes by, the top line of this list of alarms is forced off the bottom of the screen to an ethereal page below, which can only be accessed by scrolling the screen down. Let us further imagine that each alarm indication is given the same level of relevance. These are chemical process alarms—they indicate temperature warnings, pressure fluctuations, container levels, and the like. They are not harbingers of immediate doom, simply notifications that some measurement or other has gone outside some nominal target set by the engineers and programmed into the system.

So far, we are complying with the need to measure and record the information about the processes going on in our chemical complex. But what if each alarm is of the type that activates only once? As a pressure rises above a set point the alarm activates but does not reactivate if the pressure continues to rise. Nor does it clear down if the pressure falls. We soon have a situation where all day, every day, an endless list of alarm notifications slowly scroll down the screen, watched by an operator who has become well-versed in understanding their idiosyncrasies. The pressure always goes up on a Thursday, for example; or the level in container 3 always goes over when there is a delivery. It soon comes back down again. This is not complacency: this is a competent operator making experienced judgement calls against information that is not wholly accurate or reliable. It is the conflation of what Hale and Hale described as "presented information" and "expected information" or what Rasmussen described as "rule-based" and "skill-based" knowledge. And it is almost precisely what happened at the Buncefield oil depot in Hertfordshire in 2005, causing the largest explosion in Europe since the end of the Second World War. It was not the fault of any standard being complied with, nor the information being recorded. It was the *way* that information was recorded and displayed.

Case Study—Buncefield (2005)

A series of explosions and subsequent fires destroyed the Buncefield oil storage and transfer depot. Twenty-three large fuel storage tanks were damaged, and 43 people were injured, although fortunately there were no fatalities (Buncefield Investigation Homepage, 2010). The storage depot, known as a tank farm, was operated by several companies. The site received the fuel via pipelines and stored it in the tanks prior to redistributing the fuel via pipelines and road tankers to various sites in and around London and the South West.

The incident was centred on tank 912, which was being filled in the early hours of the morning of the 11th December. From approximately 03:00, the level gauge in the tank ceased to indicate any rise in contents due to issues with the monitoring equipment. At this point the fuel was being pumped into the vessel at a rate of 550 m^3 per hour. Later investigation calculated that the tank would have been full and at the point of overflowing by 05:20. The protection device that should have shut down the filling operation failed to operate, and at 05:38 the CCTV system showed a vapour cloud forming from the escaping fuel. By 05:46, this cloud was approximately 2 m deep, and by 05:50, it had begun to flow off site. Shortly after this the pumping rate rose to 890 m^3 per hour, markedly increasing the size of the vapour cloud that then ignited around 06:00 (Buncefield Major Incident Investigation Board, 2008).

4.10 Bureaucracy blinding common sense

The often well-meaning bureaucracy of form development can sometimes be of detriment to assessing risk but can also sometimes create a blind spot to appreciating a hazard in the first place. Take a busy dockyard where there is an access chamber leading to the penstocks of a dry dock. Penstocks are the heavy shutters that open and close, allowing seawater into and out of the dock in order to float out the vessel within. They are essentially large valves. The ladder down to the penstock is in very poor condition and needs maintenance. But "maintenance" on the busy dock manager's risk register features fairly low on the priority list. Worse still would be if it's recorded as "routine maintenance," which, with the best will in the world, can sometimes be treated with a rather variable attitude in terms of importance. But working below ground is just the same as working above it—it is work at height and therefore requires all the necessary protocols, including emergency rescue, as working at height above ground would. Potentially even more so due to the possible presence of noxious gases (hydrogen sulphide is not uncommon in these situations) or the reduction in levels of oxygen.

Additionally, if maintenance to our ladder is not performed with adequate priority, it may be the case that maintenance to the penstocks themselves is delayed, or not even performed at all. These are a single point of failure in a dry dock, and were they to fail when personnel are working in the dock the outcome could be catastrophic. Assigning the wrong relevance to a risk or not understanding the relevance of its consequences can create potentially harmful outcomes in safety systems if, in addition, the wrong types of questions are posed by the system. The outcome can be that the safety system reports that all is well when, clearly, there are individual issues which, if aligned, can create a trajectory through the various layers of mitigation, inevitably leading to a potentially harmful outcome.

In this way, too, it is possible that an inappropriate safety system can "hide" failings and potential failures. Individually, causative effects may not lead to a harmful event; but *consecutively* they might well do so. It is important, therefore, that safety systems have an inherent understanding of, and ability to track through, these effects in order to provide predictive capabilities based on their *relevance* and *interdependence*. The Piper Alpha oil rig disaster was not caused for want of a safety system but predominately due to the system itself not being designed robustly enough to deal with a change of shifts between workers (Oil & Gas UK, 2008). To all intents and purposes, the safety system—which in this case included a permit to work methodology—was in place and functioning appropriately, as it had for some time. Whatever causative effects had occurred up to the disaster itself (and obviously we cannot speculate on their type or frequency), they had been adequately deflected by the control measures of the safety system. However, the introduction of a shift change during the replacement of a critical valve, which was not completed prior to the shift changing, combined with a general lack of communication at that time, created a clear path, or trajectory, for the harmful outcome to manifest.

Case Study—Piper Alpha (1988)

A routine maintenance procedure caused an explosion and resultant fire on the North Sea gas production platform Piper Alpha in 1988, causing 167 fatalities (Cullen, 1990).

Routine maintenance to remove and check a pressure safety valve on the backup condensate pump involved the removal of the valve from the system on the 6th July. Due to the complexity of the work involved in maintaining the valve, it was not planned to be replaced until the following day. Permission was sought via the existing permit to work (PTW) system to fit a blanking plate in its place to seal the pipe. Although correctly fitted, the blanking plate was designed only to prevent leakage and not cope with the load of full-pressure working. That evening the primary condensate pump failed, and without knowledge of the earlier repair to the back-up condensate pressure safety valve and fitment of the temporary blanking plate, the operator started the backup condensate pump. The pressure built up and the blanking plate failed, causing a high-pressure gas leak. This quickly found an ignition source and exploded, causing a pressure wave that destroyed the firewalls intentionally designed to prevent the spread of fire. Without the protection of the firewalls, the oil pipes that ran close by were destroyed, and large quantities of oil were ignited. The built-in water deluge fire-fighting system was out of action and therefore could not be activated.

Approximately 20 minutes later, the fire was hot enough to destroy the gas pipelines fed from the other platforms nearby. These were large diameter pipes with inflammable gas at a pressure of 2,000 psi, thus significantly increasing the ferocity of the fire and causing further explosions. Most of the personnel on board the platform made their way to the accommodation area to seek safety. However, as this area had not been smoke-proofed, the repeated opening and closing of doors as they entered caused the smoke to spread further. Many of the survivors managed to escape by jumping into the sea—something they had been specifically trained not to do. The whole catastrophic incident only took 22 minutes to occur from the initial plate failure (Oil & Gas UK, 2008).

One of the issues of the bureaucracy surrounding safety—both in its implementation and management—is how administrative systems of any form can become self-fulfilling: that is to say, they become ever-more complex and bureaucratic through no more than *their need to be so*. In the early, lean stages of an organization's development, several administrative functions might be undertaken by a single individual. As the organization grows, additional individuals are employed to perform these duties and, after further growth, each individual begins to accumulate additional administrative staff to assist them. The functions of health and safety, human resources, and quality are classic examples of the sort of departments that can burgeon with the growth of any organization. Which departments grow in this manner, and the rate at which they grow, will very much depend on the organization, its primary function, and, of course, its culture. The principal outcome of this sort of administrative increase, however, is invariably that each department becomes ever more self-reliant, less communicative with other departments, and possibly self-righteous about its relative importance to the organization as a whole.

From the latter, we tend to see that systems of auditing and self-checking can become ingrained in presenting a wholesome picture of the department: no department head wants to see their department being shown in a bad light in comparison to others. This is where independent auditing, such as that which might be required by an ISO standard, for example, can be beneficial. The contrary view is, as we have discussed, that an organization that is working hard to comply with the information that an accredited standard requires can sometimes miss the importance and relevance of that information.

The propensity for organizations to behave in this way was identified by Cyril Northcote Parkinson in an essay for The Economist (Parkinson, 1955). In it, he described how administrative departments grew under their own volition, irrespective of the type or quantity of work needing to be done. As an example, he drew attention to the increase in officials at the Royal Navy Admiralty between 1914 and 1928. This showed that despite a decrease during this time of capital ships by some two-thirds, combined with a decrease of officers and men of nearly a third, there had been an *increase* in administrative personnel of nearly 80%. Whilst his observations, which are known as Parkinson's Law, were mainly directed at government departments, they are readily applicable to any organization. The internecine rivalry that can result (linked also to the fomenting of "silo thinking", where disparate departments fail to communicate or collaborate on any meaningful level) is one of the human characteristics that can affect the quality of safety systems. Systems that can be developed for the aggrandizement of the individuals concerned rather than the core purpose of preventing harm.

It is important to distinguish that this aggrandizement by individuals (or their relevant departments) of which we speak is the kind that stems from inherent human behaviour, rather than from some malicious purpose. The reasons for failures that we wish to avoid, and those of historical events that we have referenced in this book as examples, are all a result of failings in the system rather than predetermined infractions undertaken by individuals. The failings in the systems may also, of course, lead to individual failings—those that Rasmussen labelled "errors" and "violations". A properly robust safety system should take account of the potentiality of these occurring and have suitable mitigations or controls in place to deal with them.

4.11 Technology—friend or foe?

Sometimes we create elements of safety systems, such as permits to work, or, indeed, even entire systems themselves, which are supremely intricate and, we believe, highly appropriate. The subject of risk management has a long history and many of the great protagonists of the subject—Maslow, Reason, Hale, Rasmussen, et al.—introduced theories and models that we still use today. And we constantly add to these with discussions about zero-harm, constant improvement, and antifragility to name but a few. This is to be appreciated as part of our modern world's ever-growing reliance on technology and the solutions that it can provide. But we should be wary of who is providing the technology and for what reason. Systems of safety are there, or at least should be there, to prevent harm. Their design, structure, type, and methodology are all irrelevant *so long as they work appropriately*. In connection with this is one theory that perhaps gets less publicity than it should. The 14th-century philosopher, William of Ockham, postulated the theory that roughly translates as "entities are not to be multiplied beyond necessity". In philosophical arguments this principle, which we call Occam's Razor, is that when considering two competing theories, the simpler of the two is to be preferred. It was an argument later used by Galileo in his hypothesis of the heavens and is even conjured up, if considerably less eloquently, in the acronym KISS (for "keep it simple, stupid"), a design principle noted by the US Navy in 1960 (Dalzell, 2009). We may be dealing with safe operation rather than ethereal philosophical ideas, but the principle of retaining simplicity in our systems is as relevant today as ever. The reason for this is that the humans who interact with our systems today are the same types of humans who have existed for thousands of years. Apart from language, William of Ockham would behave in the same way were he alive today: he would interact with our systems (after a little training, perhaps) in the same way as modern-day operatives. Increasing the complexity of systems does not substantiate that we are more technologically minded but that we have become more *technologically dependent*. The empowerment of ensuring the correct function of a safety system is, we would suggest, therefore supplanted from the operator to the system's designer, who may be wholly disconnected from the environment in which the system is meant to be used.

Clearly, not all safety systems that have witnessed an increase in complexity and technological reliance are to be dismissed. Motor cars, for example, are today far safer for the occupants than ever before. Things such as anti-lock brakes, traction control, pyrotechnic seatbelts, and advanced tyre compounds have done much to reduce fatalities and serious injuries on roads all over the world. These devices and treatments are entirely autonomous—they do not require input from the occupants of the vehicle, although some may require maintenance by trained technicians. But why are there still so many accidents on our roads? Human error is obviously a major component that we can never ignore, as Rasmussen elegantly detailed (Rasmussen, 1982). The fact that self-driving cars in the United States, which have completed hundreds of thousands of miles worth of testing, have only been involved in a very few minor accidents lends some weight to the "human factor" being a major causal factor. But added to error we should consider complacency. Drivers who are overly confident in their vehicle's inherent ability to steer or brake its way out of trouble may become distanced from the relevance of risks that they and their vehicle face. Misplaced trust in the systems that are designed to keep us safe does, as we have stated, transfer the empowerment to maintain that safety from the operator to the designer. The operator becomes dislocated from the very premise of the safety ambition.

This is not, it should be said, the same as designing something to be safe, or "safe to operate"; a principle we discussed in *An Effective Strategy for the Safe Design in Construction and Engineering*. The creation of products and systems that are safe for their respective operators to use is a fundamental requirement for all designers, and this demands of the designer an innate appreciation of the environment and method of use. This applies equally to the design of safety systems that are required to oversee the *operation* of those products.

Having assumed that any system has taken appropriate steps to eliminate unauthorized access to it and any functions that it controls, and dismissing fears of malign intent, we must assure ourselves that any system promoting safety—be it design, production, or use—is appropriate with regard to three important factors:

- The product or undertaking in question
- The environment of its use or deployment—be that the actual environment, or marketplace, or culture, for example
- The human factors that will undoubtedly impinge upon it

With reference to the last of these factors, we should be careful not to assume "human error" to be the dark cloud of unpredictability: where there is a lack of competence;

there must always be a remedy in terms of the adequate provision of information, instruction, training, and supervision. Identifying, and making provision for, the various human factors that affect every organization should not come as a surprising nor an exhaustive task. We should be well aware of any specific factors that affect our own organizations: a high percentage of low-skilled workers; a largely peripatetic workforce; an abundance of managers compared to shopfloor personnel; a small but obstinate number of time-served staff from the earliest days of the organization. All of these are examples of the types of factors that we would be aware of and that exist—rightly or wrongly—in any organization. Correctly identifying and understanding how these factors might affect safety in any context is the first crucial component of then creating and implementing the proper control measures to mitigate them. Control measures which may, of course, include longer-term solutions such as changing the way the undertaking is performed and by whom.

5 Demonstrating "safe"

5.1 Understanding the effort required

The practice of assessing risk has been long understood, and the use of tools such as fault tree and event tree analyses has long been the mainstay of identifying where and how something can go wrong in a process, and what the likely outcome of that might be. They can also identify single points of failure which are, obviously, to be avoided wherever possible. Our understanding of the principles of "layering" control measures to provide ever-greater levels of mitigation is borne out by the ubiquitous "Swiss Cheese" model. But the bureaucracy of compliance can shift this reliance on layering through the simplification of the measures themselves. Not because of some malign intent, but rather due to a fundamental lack of understanding of the vectors of risk, such as likelihood, severity, and velocity—something that risk professionals should be well versed in.

It can be difficult to replicate the complexities of thorough risk analysis, therefore, when using predetermined forms or software. There are, of course, situations where these systems work well. An assessment under the Health and Safety (Display Screen Equipment Regulations) 1992 (DSE, 1992), for example, requires several straightforward observations to be made. Is the individual seated correctly; is the monitor adjusted correctly; is the chair adjustable for height; is there a requirement to support the feet, and so forth. The regulations provide several example measurements, both linear and angular, on which the assessor can base their assessment. In this example, the use of multiple-choice questions or drop-down boxes provides a sensible, pragmatic approach that can be replicated without the need for in-depth training. If we consider the potential complexities of something like a chemical process, the relative ease of use of this type of system may hide (or, at least, be less able to determine) more complicated issues.

Accident investigation procedures can include the "5 whys" method (Serrat, 2017) which establishes a systematic approach to pinning down a root cause. An accident has occurred: why did it happen? It happened because of X. Why did X happen? And so on. With our knowledge of fault tree analysis, and the modelling of events determined by James Reason and others, we can see that the post-event "why?" is replaced with the pre-event "what if?" This is the fundamental of risk assessing. With our computer operator and their assessment of display screen equipment, the answer to the "what if" question of "what if the monitor is not adjusted properly" is readily deduced. The operator will most likely develop musculoskeletal symptoms, most probably in their neck.

DOI: 10.1201/9781003296928-5

Failure Mode Effects Analysis (FMEA) and Failure Mode Effects and Criticality Analysis (FMECA), were tools used to identify potential failure modes in a defined system (NASA, 2012). FMECA was an extension of FMEA whereby the criticality levels were used to prioritize and manage the effects. The tool was applied in two stages, firstly using it to identify all the possible failure modes associated with the system and the potential effects that they may cause. Then secondly to critically analyse the ranking of failure modes according to the probability and severity associated with each failure mode.

5.2 Putting the effort in perspective

Ever since mankind started to design and create products for his work, protection, and edification, the records have been filled with examples of those that per-formed marvellously—as well as those that failed dismally. The ability of Concorde's engine cowls to reduce the intake airspeed from 2,200 km/h to around 800 km/h in just under 3.5 m and the failure of the Tacoma Narrows (Advisory Board on the Investigation of Suspension Bridges, 1944) bridge are but two examples of each type. We rarely hear about the middle ground of so-so products, however, that fulfilled their requirements with aplomb but without distinction. We seldom think about the "also rans". Perhaps the reason for their success in obscurity is that, by and large, these "also rans" managed to hit the sweet spot of proper, safe design coupled with appropriate, safe operation. The failure of a bridge, aircraft, or style of mobile phone handsets might make national headlines, but the vast, dowdy middle ground of func-tionality doesn't make us salivate with schadenfreude.

An alternative, if somewhat speculative view, is that luck has played a part in ensuring that the holes in the layers of mitigations and control measures have never *quite* aligned up during these products' life cycles in order to provide a clear path for failure to make its way from threat to harm. Certainly there have been instances where, but for the grace of providence, some product or other would have suffered from a harmful event were it not for some kind of intervention. An inspection engineer discovering a stress fracture; a repair technician discovering a loose nut; an apprentice designer re-checking the wind load data on a skyscraper, for example. But we should consider that the vast majority of products that have ever been designed and constructed (or made, if you prefer: some people cannot help thinking of bricks and mortar when they hear the word "construction") have, in fact, been successful—in terms of safety—because things were done *properly and appropriately*. (In essence, this is the dilemma of anyone practising in safety: the prevention of things going suddenly and spectacularly wrong is our entire raison d'être. The output of our purpose is, therefore, continued serene peacefulness; which is hardly exciting, or even noticeable. The best thing any safety professional can say at the end of each day is, "nothing happened". It's not an easy sell.)

There are, quintessentially, two main areas that create the foundations for things to go wrong in any project or system that delivers a product into the world: finances and communication. We shall not be dealing with finances in this book per se; we shall instead be assuming that adequate financial resources are available at any given point. We do, however, advocate the proper and effective deployment of finances and will, moreover, demonstrate how the effective intervention of safety management can save money in the long run. The reason we are not considering financial resources as a source of failure is quite simple: under UK law it is illegal to scrimp on the availability of financial resources in relation to safety.

Health and Safety at Work etc. Act 1974

Section 2 requires the provision and maintenance of plant, systems, arrangement for safety, information, instruction, training, and supervision appropriate to the undertaking to ensure the safety of employees, so far as reasonably practicable. Section 37 allows for an officer of a body corporate, as well as the body corporate itself, to be prosecuted where there is consent, connivance or neglect proven in the commission of an offence.

Management of Health and Safety at Work Regulations 1999

Regulation 5 requires arrangements for the effective planning, organization, control, monitoring, and review of preventative and protective measures to be implemented, with due regard to the organization's size and nature of undertaking.

Sentencing guidelines for health and safety, corporate manslaughter, and food safety offences

Step two of the guidelines requires magistrates and judges to consider factors which increase the seriousness of any offence, the first of which listed is "cost-cutting at the expense of safety".

The deliberate withholding of adequate resources from safety-related purposes is legally, as well as ethically and morally, wrong. Even where safety-related matters during design, manufacture, or operation are curtailed due to unforeseen financial restrictions—such as a project running out of money, for example—there should be the organizational will to either procure additional funding or stop the project (or operation) until such time as the appropriate resources are available.

Communication—or the lack of it—is, therefore, our main focus in portraying what can go well, and not so well, in maintaining safety. In our previous book, we discussed how, during the design stage of a product's life cycle, it is important that the operators, maintainers, and repairers of any product are involved in that design. This has the benefit of ensuring that operational safety-related matters which the designer may not be aware of are discussed at the early stages of design. It is always easier—and less costly!—to amend a drawing than a completed product. We also discussed how, in usual practice, the engineers and designers working on a design are often disconnected from the safety practitioners who have the task of making sure the end product is operated safely. This is due to a number of professional, functional, socio-political, and sociocultural reasons we shall not elaborate on here: suffice it to say that this disconnect can be deleterious to safety, and safe operation, throughout any product's life cycle. In unemotional analysis, however, it makes no sense that those who are tasked with identifying the risks in producing something—the designers and engineers—should be dislocated from those who are tasked with ensuring that it remains safe during its use. After all, safe operation—and by extension, safety itself—is the product of the risk management process.

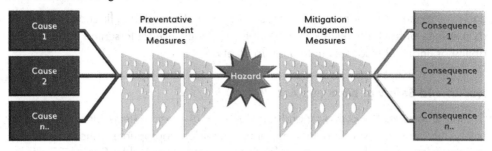

Figure 5.1 Adaptation of the bow tie methodology.

To demonstrate this, we should turn to the bow tie methodology of risk identification and management (see Figure 5.1).

Ostensibly, the bow tie consists of preventative measures on the left-hand side and mitigative measures on the right. In the centre is the "event". This could be considered an outcome from some realized risk, or it could be a hazard in its own right, whereby the fault tree deals with possible failings of that hazard and the event tree deals with possible risk outcomes. We might also view the bow tie as being concerned with *risk* on the left-hand side and *safety* on the right. This is relatively straightforward if we imagine the centre of the bow tie as a hazard. The preventative measures are concerned with the probabilities of failures should certain conditions be realized. It allows engineers and designers to predict where failure points are likely to be and, most importantly, where any *single points* of failure occur. From this analysis, we can see if welds should be strengthened, valves duplicated, sensors provided with redundancy, and so forth.

The mitigation measures are concerned with what happens after the event, or in this case, what possible outcomes could be realized from a hazard. A hazard may have several possible risks attached to it, and we will undoubtedly have a number of control measures in place to deal with them. The event tree looks at how harm can track through these layers before ultimately becoming a harmful outcome. By prescribing values to the various factors and layers in both the fault and event trees, we can determine probabilistic outcomes. The layers we determine in the event tree analysis can be used to help us review the risk assessment that we develop for the use or operation of the hazard. It would be useful here to distinguish between the *quantitative* methodology usually used in processes like fault and event tree analysis, and the *qualitative* methodology generally found in risk assessments. Using numbers in analysis models allows engineers to predict events with some certainty, but the critical detail is obviously the numbers themselves. Accepting that a valve for example has a 1-in-a-100 chance of failure is to accept that this was determined using the proper critical analysis originally; and that the valves were made according to the production specifications demanded by that analysis. It is also in part to ignore the words of Sir Terry Pratchett when he suggested that million-to-one chances tend to crop up nine times out of ten (Pratchett, 1987). Conversely, qualitative assessments require skilful understanding of the input data and the reasoning behind it, as well as an accurate assessment of the output or findings.

The bow tie model clearly demonstrates the cohesive link between the safe design, manufacture, and use of any product. That product in this case is the hazard, which could be an office block, jet fighter or sausage-making machine, or any component part of any one of them. The designers or engineers concerned with the left-hand side of our bow tie would use a safety case to demonstrate the mitigation factors that they have instigated into the design. The manufacturers of the product would create a validation plan to demonstrate that the product has been made in accordance with the specifications designated in that safety case. And the user or operator would have a safe system of work to ensure the product is used in accordance with the designer's intent and the manufacturer's instructions—both of which stem from the safety case and the validation plan, respectively.

The safe system of work, it should be remembered, is not a "thing" in its own right. It should consist of the following to whatever level or extent is required or relevant:

- Job safety analysis—the work is broken down into its constituent parts from start to end, usually as a long itemized list.
- Risk assessment—identifying the risks at each stage of the job safety analysis, and who might be harmed by them.
- Work instruction—this might cover who is authorized to operate or use, maintain, and repair the product; any specific training or supervision that is required; any action plans to deal with exceptional use such as emergencies, or abnormal use such as maintenance.

We could say that the designers and engineers (the left-hand side of the bow tie) are concerned with *risk management*, and that the operators, or the safety practitioners who guide them (the right-hand side) are concerned with *safety management*. What we are sure of, however, is that both sides are concerned with the *prevention of harmful outcomes*. Both use diagnostic tools; both use foresight, intuition and experience; and both use methodologies to reach an appropriate conclusion. The terminology may be different, but the intentions are very much the same.

5.3 Similarities between safety case and golden thread

In the wake of the Grenfell Tower tragedy in 2017, the UK Government commissioned a review of building regulations and fire safety, chaired by Dame Judith Hackitt. The review was called "Building a Safer Future" (Hackitt, 2018) and was published in May 2018—curiously, some 17 months before the first phase of the public inquiry report into the entire event, led by Sir Martin Moore-Bick. The review concluded that further regulation was required in the construction industry, particularly in that area connected with what it termed "higher risk residential buildings" (HRRBs) which were later defined as those above 18 m. The public inquiry, however, began to reveal that some of the causes leading up to the fire at Grenfell Tower were due to a lack of compliance with existing regulations and standards and had even involved some connivance or malfeasance by those involved in supplying the work to refurbish the tower between 2012 and 2016. It also noted the failure of enforcing authorities to apply the lessons that had been learned from the Lakanal House fire (Knight, 2009).

Clearly, the introduction of additional regulation is of little value where there appears to have been a concerted effort to wilfully ignore already existing regulations.

The Building a Safer Future review had, as one of its key conclusions, the introduction of a "golden thread" of information, to be digitally based and which would run throughout the building's life cycle to ensure that the design intent of the original design was maintained. The review recommended that this thread would begin with the Building Information Modelling (BIM, 2012) approach. This approach has long been used in engineering construction sectors, such as aerospace and automotive manufacture, and was declared a requirement by the UK government on centrally procured projects since April 2016. Unfortunately, like the requirement for the excellent PRINCE2 management methodology (What is PRINCE2, 2022), the application of BIM to any project comes at an additional expense which is not always something that the client is prepared to bear. BIM in traditional construction has, like PRINCE2, become a "nice to have" feature—even in centrally funded projects—despite the best of intentions, primarily because of the cost but also because there has not been the will to enforce its use. The outcome of not adhering to basic regulation, let alone best practice, is that design failures, connivance—and ultimately, tragedy—will continue to occur.

But BIM alone is not the whole answer. The Building a Safer Future Review noted some "recent notable examples" of projects that had applied BIM, which are listed below. We have noted against each one why the use of BIM in these notable projects may not have been considered an overwhelming success.

- Heathrow Terminal 5—During its first five days of operation in March 2008, the airline at Terminal 5 lost 23,000 items of baggage, cancelled 500 flights, and made losses of £16 million due to myriad issues with the building and its infrastructure.
- Ministry of Justice prisons—Newly built so-called mega prisons, such as HMP Berwyn, have issues such as facilities not built to standard, extremely poor acoustics, exceptional delays to fit out of certain areas, and lack of facilities for prison staff.
- Cross Rail—One of the largest construction projects ever undertaken in the UK, Cross Rail is, at the time of writing, some three years behind schedule, around £5 billion over budget, and without a guaranteed full completion date.

But we should not dwell on the few obvious exceptions: building information modelling is an excellent method for ensuring that a design meets with the expectations of the client's intent and can form the first part of the golden thread of information that will inform all those who have to construct, operate, maintain and finally dispose of any product. The important factor is that of compliance with the *intention* of BIM, or a safety case, or of a health and safety file—or whatever new term is used for essentially the same thing. In the Building a Safer Future review, it states that:

Success in implementation will depend upon effective leadership and collaboration, as well as ensuring sufficient resources across the industry. There must also be a strong commitment from industry to improve the standard of records kept and to ensure their maintenance for both new and existing buildings.

This is a fine intention, but it cannot be left to the construction industry—nor indeed, any industry—to police its own compliance. The observations of the Grenfell Inquiry would seem to support this. Similarly, we must not presume to ensure compliance with safety law simply because to do so is legally circumspect. Viscount Halifax (1633–1695) wrote, "men are not hanged because they steal horses, but that horses may not be stolen" (Ratcliffe, 2016), meaning that we do not have laws for the sake of providing an outcome, but in order to prevent societally unacceptable things from happening. Equally, we should not have laws and standards regarding safety for the sake of just having laws, but because the prevention of harm should be our fundamental determination in any design and operation. The laws, the standards, the guidance, and even the ambitions of the sentencing courts, are all currently in place to persuade us to achieve the highest standards in our duty of care.

For many years, engineers of all persuasions have used the safety case to demonstrate, at the design phase, their overall compliance with the design intent and to demonstrate the levels and factors of safety that have been applied. The main elements of the safety case are as follows:

- Requirements
- Claims
- Evidence
- Arguments
- Inference

The claims that a design is safe to operate are based on documented evidence that can form part of a structured argument. This demonstrates that hazards within the design have been identified and that risks deriving from them have been reduced to a level as low as reasonably practicable (ALARP). This is sometimes known as "making an ALARP statement". The identification of risks that remain is an output of the safety case in the form of operational or limitational instructions. A safe working load or maximum rotational speed are simple examples of this sort of instructional information. All too often, the safety case is not widely recognized beyond the designers and engineers who develop and create the product in the first place; and yet, this is precisely the sort of information that can be used to develop the preliminary information for the golden thread that is now being fashionably discussed. The safety case can be used to prove design risk assessments; form the base arguments for a building information model; and, qualify the required contents of a health and safety or technical file. (Health and safety files and technical files form part of the evidence required by a safety case, therefore, both can be checked back to the safety case to ensure they contain the relevant information that was required as part of that evidence.)

Howsoever the information pertinent to all these methodologies and their relevant laws and standards is routed, there are perhaps two fundamental issues which need addressing: the format of the information and the will to complete it. The Building a Safer Future review concludes that a digital format is the correct course and, in our world of almost exclusively digital communication, this is perhaps a fair assumption. Interestingly, the coroner's inquiry into the 2009 Lakanal House fire made a recommendation in 2013 to "commence storing CDM H&S files electronically ..." (Kirkham, 2013). But digital communication is a fickle gamester in that it is predominately led by a very small number of very large corporations whose ambitions are not the betterment of mankind but the enlargement of their cash reserves. Protocols for the exchange of digital information are based on the vagaries and fashions of a small number of software designers rather than the needs of the vast number of people who use their software. What can be read and stored with ease today may not be so easily accessed in the years to come. Anyone who remembers magnetic tape, floppy drives, Zip drives, DAT tape, and even CDRs will understand this problem. If we imagine a product—an office block, say— with a projected 50-year lifespan and then recall the technology for storing information from fifty years ago it is easy to see that several revisions of the storage medium would have had to have taken place, let alone the innumerable revisions to the software protocols. It is not without good reason that the UK Government still to this day has new laws written on vellum and stored in a controlled environment. Ink writing on vellum will last many hundreds of years, notwithstanding the evolution of the syntax of the English language.

The importance of adopting and using a system, therefore, that can be adapted through the years of any product's life cycle is linked, then, to the second point of having the will and the wherewithal to do so. The documentation system must be adaptable, updateable, transmissible, and accessible to everyone who has a need or requirement to interact with it. It must be able to be updated by potentially several different document owners throughout the lifetime of the product, with each successive owner taking responsibility for the documentary system's integrity. This can only be achieved through a concerted effort from industry and government working together, combined with public expectation. The aeronautical industry, for example, has long had its own equivalent of the "golden thread" of information throughout an aircraft's life cycle. This has been due to a combination of aeronautical engineers' preoccupation with safety, governmental influence in terms of regulation and licensing, and the public's obvious concerns with their own personal welfare when flying. Detractors may point to the large difference in profitability of some traditional engineering sectors with traditional construction, and this is perhaps where society as a whole needs to place its attention in the first instance. Product creators cannot spend more on developing their products if the buyers are not prepared to pay more for them. And all the time that creators of product see embedded safety as an on-cost rather than an inherent cost this race to the bottom will continue.

The Building a Safer Future review is predominately concerned with domestic buildings above 18 m in height, which may seem a little odd, as buildings of any height or shape can provide the necessary circumstances for a disastrous fire. It is interesting that even the chair of the review could not provide a convincing answer as to why the review was fixated on this very specific height. Certainly, it is unlikely to relate to the reach height of fire appliances—in London, for example, there are aerial appliances with up to 64 m reach. This would lead, through the review's recommendations, to a situation where only some types of construction needed a "golden thread" of information, where others would not. It should also be borne in mind that even though the Construction (Design and Management) Regulations (CDM, 2015) detail a large range of types of "construction" to which they refer (and, indeed, should do, as we argued in our previous book), the list of health and safety file contents in Appendix 4 to the regulations is very much concerned with the sort of documentation that one would expect from a "traditional" construction project: that is, any building of some description.

We discussed, too, in our previous book how this information can be correlated with the output of other types of construction projects, i.e., those that output some other form of product—an aircraft, a ship, a motor car and so forth. We also examined how the other prescribed documents of CDM could be applied successfully to any project with some form of manufactured output. Equally, we will discuss here the correlation between what the review terms a golden thread and the expectations of engineers in what they call a safety case.

The importance of the safety case, other than to gather the information and arguments that support the methodology of the design intent of a product, is to provide future owners/operators with the information necessary to ensure that any redevelopment or amendment to the product is completed within its original safety limitations. Consider an armoured vehicle used by the military. After a few years of service, it may be necessary to upgrade its armour, its weaponry, or perhaps introduce additional equipment such as for mine detection or reconnaissance. Simply drilling a hole into an armoured vehicle is clearly not a wise idea, as it could severely compromise the vehicle's integrity. The additions or amendments to the vehicle must be done sympathetically to the original design intent. This is how the British Army's Challenger 2 battle tank is being upgraded to become "Challenger 3".

The safety case demonstrates in detail how an armoured vehicle arrives at its final design: where the stress points are, the moments of force in the structure, the location of wiring and pipework, and what calculations were made in arriving at the design— and therefore what limitations there are in applying future forces and loads, etc. The safety case outlines any residual hazards as well as indicating what measures have been applied to deal with those hazards that were identified in the original design. All of this allows future engineers and designers to ensure that any upgrades or amendments are made with proper regard to the inherent safety of the original design. In short, the risks associated with future design changes will be kept *as low as reasonably practicable* (ALARP).

In essence, this is similar to the information required by the health and safety file, a documentary device required by the first CDM regulations in 1994 and still required in that legislation's present incarnation. But adherence to the requirement for the health and safety file in the traditional construction industry remains poor. Another of the recommendations made by the coroner's inquiry into the Lakanal House fire in 2009 was that there should be "access to relevant information about the design and construction of high rise residential buildings and refurbishment work carried out to enable an assessor to consider whether compartmentation is sufficient or might have been breached".

5.4 The case for safety vs. the golden thread

The current discussion centred around the need for a "golden thread" of information is driven by the need for better safety in domestic high-rise properties as a result of the Grenfell Tower fire. This is laudable, and it is certainly an important issue that the occupants of such buildings are as well-informed as possible to ensure their safety. But this requirement for such information has been around since 1994, albeit not in the particular format that is proposed in the (Building Safer) report. But the intention was certainly there, although this was perhaps never fully considered.

The proposed golden thread unfortunately also suffers from some poorly considered issues. It requires the extension of information gathered through the building modelling process (BIM) to inform the user/operator of the building of the original design intent. This should, of course, include any design limitations, as well as specification data regarding the types of materials used in the construction. This would highlight, at least to the technically minded, any potential risks posed by such materials or design concepts—especially those regarding fire safety and emergency egress. The issue here is that BIM has not been as widely engaged with as it deserved. Originally developed by the UK government as a means to reduce waste, improve safety, and reduce construction costs, it has as we have seen become something of a "nice to have" rather than an entrenched philosophy. Even central government-funded projects that the authors have been involved with have not been to BIM specifications due to the perceived increase in cost, time, and effort required to implement them.

Another issue is that the golden thread is aimed at a very narrow part of just one industry sector—construction. This is unfortunate because, as we have discussed, the provision of the right information at the right time to the right people is crucial for the delivery of safety in *any project*. If we are to propose that the construction sector is more susceptible to error and misjudgement than any other industry, then are we not conceding that nearly 30 years of construction industry-specific regulation has not yielded the results that we expected? On the other hand, if we were to enforce the construction regulations on all the disparate industries to which they *actually* apply, then we are admitting that any industry that produces something is covered by them and, therefore, should be benefitting from embedding safety from the outset of any project.

The key element missing from the golden thread, in comparison to a safety case, is the lack of verification of the information provided. In other words, the golden thread requires information regarding the materials and processes involved in a design (as is currently the case with the health and safety file requirement of the construction regulations), but it does not include a facility to verify that those materials or processes were actually implemented correctly, nor maintained, repaired, or replaced in the correct manner. This was a fundamental issue with the redevelopment work carried out at Grenfell prior to the fire. A safety case, by comparison, requires the addition of an "argument" in the equation of information (see Figure 5.2). This ostensibly provides the verification that the evidence supplied—the list of materials, processes, limitations of use, and so forth—do in fact meet the original claim. The argument might be the demonstration of how materials are maintained throughout their lifetime or how limitations of use are enforced during operation. This may include any of the factors that we define in Chapter 6: such matters as training, documentation systems, authorizations, skill sets, etc. Whilst the golden thread has received much attention of late and is certainly a welcome addition to the wider conversation on safety, it would be exponentially improved by being aligned with the pre-existing process methodology of the safety case, combined with our outline of considerations in Section 6.3. This would be, we would suggest, the proper case for safety.

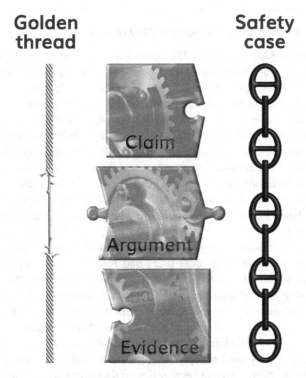

Golden thread **Safety case**

Claim

Argument

Evidence

Figure 5.2 Golden thread and safety case in comparison to the components of safety.

Again, we should appreciate that the requirement for this information, and the methodologies for assimilating it, are nothing new. What has been lacking in the past is the cohesiveness of that information: its portability and acceptability from one phase to another. And, perhaps, the acceptance that safety in any given project development is not a stop-go process but should, in fact, be one of continual and rolling progress. From this development or flow of information, we can deduce that it is perhaps not wholly possible to, firstly, design or engineer a product without a proper understanding of the entire *use* to which that product will be put. Nor secondly, is it entirely possible to assess the risks involved during the product's use phase without an understanding of how—and, indeed, why—certain elements of the product were designed or engineered in any particular way. Not that this is to suggest that the safety case, design risk assessments, and manufacturing methodology should all be presented to the end user of, say, an industrial machine in order for them to develop a suitable operational risk assessment. Nor is it to suggest that these would also be required to complete an operational manual or training programme. Indeed, these last two items should have been contemplated during the design phase itself (see the chapter on Design Process Considerations in *An Effective Strategy for Safe Design in Engineering and Construction*). Instead, we should be inclining our thoughts towards amendments, modifications, repairs, and maintenance carried out throughout the lifetime of the machine, building, or whatever.

Case Study—Royal Air Force Nimrod XV230 (2006)

A Royal Air Force (RAF) Nimrod XV230 aircraft was on a routine mission flying over Helmand Province in Southern Afghanistan when a fire started, causing a mid-air explosion killing all 12 service personnel on board (Haddon-Cave, 2009).

The Nimrod took off from its operational base and, in order to complete its extended mission, was refuelled in-flight from a Tristar tanker, which was a well-practiced procedure. After refuelling, two warning alarms were activated, indicating a fire in the bomb bay and the presence of either smoke or hydraulic mist in the elevator bay. The aircraft depressurized, and the camera operator reported flames coming from the rear of the engines on the starboard side. The crew carried out emergency drills, transmitted a Mayday call, and diverted course to a nearby airfield. It was spotted by an RAF pilot who reported flames trailing from the starboard wing just before it exploded 3,000 feet above the ground.

The cause of the disaster was associated with two specific design modifications—the Supplementary Conditioning Packs (SCP) and Air-to-Air Refuelling (AAR). The SCP design required a feed-off from the engines' hot air exhausts to pass through the No.7 dry tank. The AAR modifications required fuel pipes to run through No.7 dry tank. The AAR modification also required special blow-off plates to be fitted to the fuel tanks to allow any excess pressure in the fuel tanks to be relieved to atmosphere.

Most of the approximately four-hundred fuel couplings on the Nimrod were of a type that had been previously used on Spitfires, Lancasters, and Vulcan bombers. The "O" seals used in these fittings were only ever replaced during corrective maintenance when leaking was observed. The remainder were of a type which were known to have been made out of non-conforming materials after the year 2000.

The insulation on the hot air exhaust pipes was known to deteriorate, and, during the upgrade process, degradation to pipework connectors was also noticed. The vent pipe was covered in a non-absorbent covering to prevent fuel from soaking in. Flexible muffs, laced up with steel wire, were used to cover expansion joints, although one of the muffs fitted on the SCP did not cover the hot elements as effectively as it should have.

Even though there was both a fuel and ignition source present in the No.7 dry tank, the area was not designated as a fire zone, and there was no fire detection or suppression system in the tank.

It was not known exactly which of the coupling types failed, but there was a significant fuel leak into the No.7 Dry Tank which was ignited when it came into contact with the hot air exhaust system, resulting in a catastrophic explosion.

5.5 Components of safe

Almost from the birth of the railways, with the first Act of Parliament granting permission for the Liverpool and Manchester Railway in 1826, the ability to operate trains profitably and safely was paramount. The introduction of the "block" system provided a means to greatly improve line traffic but did little for improving safety. The earliest signalling system consisted of "policemen" with stopwatches at the side of the track, positioned at strategic locations, who timed passing trains in order to indicate to approaching trains how far in front, in terms of time, another train was. This allowed increased traffic but resulted in many accidents, which did little for the reputation of this fledgling technology. By 1844, there were around 2,000 miles of track in Great Britain, and the railways were having a serious and irreversible impact on the way that people all over the country were living their lives and conducting their business.

The introduction of semaphore signalling in the 1840s allowed for further increases to line capacity whilst also creating a much safer environment for trains to operate in. Via a system of communication between signal boxes, train drivers were now able to be informed of the condition of *distant* signals; i.e., those at the head of the next "block". But the real benefit of the system, from a safety perspective, was that the controlling mechanisms in the signal boxes were interlocked to ensure safe operation. Complex and exquisitely engineered interlocks prevented a signalman from allowing a train to pass one signal if another that was pertinent to the block it was due to travel on was not set correctly. In addition, semaphore signals denoted "stop" when they hung down at 45 degrees (later signals known as upper quadrant denoted "stop" when horizontal). This meant that, in the event of a break in the linkage from the signal to the signal box, the semaphore would naturally fall to the "stop" position in order to prevent the train passing until the problem had been rectified. This was an early adoption of the fail-safe principle. Later transition to coloured light signals may have improved many aspects of signalling safety, but the basic principles of the system created over a hundred years ago (i.e., those of maximizing line traffic whilst maintaining safe running) remain.

The upshot of all this is that the signalling system devised for the fledgling railway network of Great Britain was one principally of simplicity. It had a strict and easily trained protocol, and it had designed-in safety features. Notwithstanding the large maintenance regime required, the system was extremely straightforward for such a relatively complex and potentially high-risk operation. The system did still rely on human interaction and, therefore, the possibility of human error. However, the introduction of train protection warning systems and, later, automatic train protection systems helped to reduce this possibility. But as much as technologists might argue that this demonstrates how technology can eradicate human intervention (and thereby, human error), technology is not the only answer. This is because any form of technology designed to perform some function or other is the product itself of myriad inputs that created it in the first place. And those inputs are, by default, themselves a product of the culture that spawned them. Just as equally, that culture is the evolution of the demands, needs, and ambitions of the originator of the statement of requirements that demands the new technology. In short: a culture that thrives on technology—let's say, an organization that sells technology to the masses—is more prone to determining that technology is the answer to its organizational problems instead of, perhaps, better training, welfare, or pay.

To elucidate: we may feel justifiably righteous about the amount of technology that shapes our lives today, but was all of it designed with the benefit of mankind in mind? Is not some of the ingrained, perhaps overwhelming technology that we have—from smartphones to social media—not just a creation of some organization's or individual's desire to make a voluminous amount of profit? Equally, the various improvements to the loom made in the late 18th and early 19th centuries were not made ostensibly to improve the safety or health of the workers that used them but to improve productivity. More output means more money. That some of the improvements meant fewer people were required to use the machines, meaning fewer people were exposed to their dangers, was perhaps just a by-product; albeit a valuable one.

Therefore, in turning to any form of technological or engineering solution, we must ensure that the reasoning behind that solution is empathetic to our need to improve safety if safety is the thing that needs improving. Similarly, if an improvement in safety is not the driver for change, then we must ensure that any solution has, at worst, no detrimental effect on safety whatsoever and, at best, improves it to some degree.

This is clearly not the same as performing a risk assessment but is not wholly disconnected from it. Indeed, the considerations here can be very much used to support and inform a risk assessment conducted on the shop floor. It is in the layering of the detail that this process of consideration varies. In a standard risk assessment, the emphasis is very much on what a particular task is, how it is performed, and who might be affected by it and how. Thus we can understand the *immediate* risks involved in the task. The process of assessing the safety-related elements surrounding that task allows us to see if particular events or conditions will have a potentially negative impact on it. These are the *inherent* risks involved in the task. Examining two high-profile events, both occurring in the high-risk oil extraction industry, will help us to see how this might be.

As we illustrated earlier, the series of explosions on the oil rig Piper Alpha were as a result of routine maintenance. Although carried out under a permit to work system, this did not factor in the transfer of information when there was a shift change. Specific training on the system was also subsequently found to be poor. As no proper handover was conducted between the shifts, the need to bring the temporarily repaired pump online caused the catastrophic release of gas and its subsequent ignition. The investigation that followed discovered that there had been a fatal accident on the same platform only the year before due to a permit to work not being issued. In standard risk assessment terms, we could surmise that this previous accident—which resulted, incidentally, in a prosecution—improved matters because a permit to work system had been adopted and was in place when, a year later, the dayshift removed the fateful safety valve.

In 2010 there was similarly a series of explosions on the Deepwater Horizon oil rig, operating this time in the Gulf of Mexico.

Case Study—Deepwater Horizon (2010)

The BP drilling rig Deepwater Horizon suffered a massive explosion, destroying the rig and causing 11 fatalities among the crew members and seriously injuring a further 17 (BP, 2010).

The rig was floating in nearly 5,000 feet of water in the Gulf of Mexico, and planned work to extract oil was behind schedule. On 20 April, during the daily operations conference call, a team of specialists was summoned to fly out and perform some tests on the bottom cement seal that had been poured the day before. The cement used was a special light nitrified foam slurry designed for use at this depth and pressure. As the cementing appeared to have gone so well, it was decided that the specialists would not be required to run their full set of tests, thus saving time and, coincidentally, some $128,000 in fees. Other tests, such as the positive and negative pressure tests, were then carried out by the drill rig team in order to check that the well head was robust enough to remain in place until, eventually, the rig had extracted all the viable oil and gas from the site and was decamped to another location, a term the industry call abandonment.

At around 17:00 on the evening of the 20th there were some issues with the negative pressure test, which did not get reported in the confusion caused by an excessive number of people in the control room during the watch changeover. The negative pressure test was subsequently performed using an alternative method which convinced the team all was well, and so the blowout preventer was reopened, and sea water began to be pumped down the drill pipe to displace the mud and spacer. Just after 21:00, a high frequency vibration was detected by the bridge watch keeper, who went outside and saw that a nearby supply vessel was covered in a muddy film. The vibration was followed by a loud hissing noise.

High volumes of mud and gas emerged from the degasser, and the first explosion occurred. All emergency functions built-in to control the well, and in this case designed to shut the well down to prevent any further release of hydrocarbons, failed.

The operation of oil rigs is one of the most hazardous undertakings anywhere in the world. The risks to safety, considering the hazards and environment in which many of these rigs operate, can be huge; not only in terms of the level of harm but also the number of individuals at risk of harm. There are also of course the potentially huge risks to reputation, productivity, and asset damage, and not least the possible environmental damage which, in the case of Deepwater Horizon, was devastating. These types of operations, as with so many others of all levels of hazardousness, are generally well-versed in the assessment of risk and the implementation and management of control measures. Many of the great strides forward in the provision of safety over the last few decades have been due to legal or mandatory requirements, as well as society's ever-increasing abhorrence generally to safety-related disasters at places of work. But we know as engineers, designers, and risk and safety professionals that control measures form in layers, and faults in those layers of controls can "align" from time-to-time. It is at the conjunction of these alignments that we can be exposed to the greatest risk in the system, regardless of how well we might believe we have assessed the immediate risks to it. These layers of control—their function, management, and the wherewithal to maintain them—are precisely the sort of inherent risks to which, organizationally, we should turn our attention.

In the HSE's guidance for investigating accidents (HSG245, 2004), it states:

> *Adverse events have many causes. What may appear to be bad luck (being in the wrong place at the wrong time) can, on analysis, be seen as a chain of failures and errors that lead almost inevitably to the adverse event.*

This "chain of failures" is another way of describing the faults in the layers of control measures aligning so that the risk of harm is realized. It is also known as (and is represented in the HSE's guidance as) the domino effect. We have seen how these faults, or falling dominoes, aligned in the examples given above to devastating effect, despite there being multiple safety protocols in place.

But the faults described here happened in relatively short succession. Sometimes it can take many years for these faults in the control measures to align or for the dominoes to fall. The I-35W bridge collapse in Minneapolis is one example of this (FEMA, 2007). Completed in 1967, the bridge was modified in 1977 when the road deck thickness was increased, and again in 1998 when barriers and a de-icing system were installed. Together, these modifications added around 1,900 tonnes of additional mass to the structure. Over forty years of use, the bridge developed several areas of corrosion and cracking, but these were not individually of major concern and continued to be monitored as part of the regular inspections that took place on a more frequent basis than that advised by the highways authority.

Of particular note, however, was that some of the gusset plates used to connect the trusses of the structure were not of the thickness required in the design. Plates of 35 mm, 25 mm, and 16 mm were used variously on the structure, but some were only 13 mm thick, and this was not corrected at the time of construction, nor was it reported during any of the periodic inspections even though it had been identified at least once by an inspection engineer. One of these thinner plates, identified as U10W, was also showing signs of buckling, and this issue too was not reported. This buckling indicated a shearing force between the two truss members that the gusset plate was secured to. On 1 August 2007, some maintenance work was being carried out on the bridge which involved an amount of aggregate being piled-up on the road deck along with various construction plant and vehicles necessary to complete the work. All this construction-related machinery and material, whose combined mass was around 260 tonnes, was located around the area of the gusset plate U10W. During the evening rush hour, as cars slowly traversed the bridge—the number of lanes available having been reduced due to the maintenance work—this plate gave way and the bridge catastrophically collapsed.

What is telling about the situation with the I35W bridge, although regrettably not unusual, is that several layers of control measures were in place, but because they all had inherent faults, there was always an opportunity for them to align and cause failure. Table 5.1 highlights possible faults with some of these controls.

Table 5.1 Examples of possible faults with control measures

Consideration	Control measure(s)	Fault
Design	Engineering standards applicable to structures	Engineering formulae used to determine the correct strength of components do not always mean that that is what is designed or installed
	Building standards applicable to bridges	The correct specification of designed components does not necessarily translate into the correct components being delivered to site
Construction	Validation process to confirm compliance with design criteria	A validation process may only confirm the type, serial number, or quantity of components used rather than their absolute specification
	Sign off by client accepting the asset into service	The client may have to make assumptions that those who are responsible for the validation prior to handover are competent (indeed, this is undoubtedly what the client has paid handsomely for)
Repair	Engineering standards applicable to increased loads	Engineering formulae used to determine additional loading capacities require the condition of existing components to be known—if this information is not available or inaccurate, the formulae are redundant
	Building standards applicable to the quality of repairs	Additional load ratings (conducted on I35W in 1979, 1995, and 1997) should include information about any components whose condition is sub-standard

(Continued)

Table 5.1 Examples of possible faults with control measures *(Continued)*

Consideration	Control measure(s)	Fault
Maintenance	Highway authority standards for repairs on live assets	The mass of plant and vehicles required for works may not be known, nor might assumptions be made about the capacity of structures to accept additional or point loadings
Inspection	Code of practice for inspection engineers	Codes of practice, chartered status, or experience are not guarantees against the instance of human error, judgement, or bias
	Engineering standards for identifying inadequate components	The specifying of what is and what is not adequate in terms of component integrity does not necessarily translate into that standard being applied in the field

As with any historical incident, it is unhelpful (and possibly churlish) to suggest that with a more robust system in place, or with better tools or diagnostics, such an event would not have occurred. With the I35W bridge collapse, as with so many other similar events that have been studied, the concatenation of circumstances is likely to have occurred at some stage *even if one or other of the individual circumstances were altered.* We imagine the control measures—the layers of Swiss cheese if you like—as a linear thing, whereas, in fact, causality finds its path like water running through a dry riverbed or lightning coursing through the night sky. Risk will follow the path of least resistance; where control measures are weakest. Hence the reason for having a hierarchy of control measures; some controls will be more effective than others. And if those more-effective controls are the first to be applied, then the risk becomes ever-weakened in its efforts to challenge each successive control or layer. Efforts to eliminate, substitute, or engineer out the risk are the most effective; those that rely on an individual's interaction, choice, or judgement remain the least effective.

But this system of a hierarchy of controls also has layers of control measures within itself. The more resilient controls—elimination, substitution, engineering solutions—may be weakened by decisions made when implementing them. Elimination may simply move the risk to another location or operation. Substitution may do likewise or introduce alternative risks connected with the process or material that has been substituted for another. Engineering controls will be based on engineering principles, but these are only effective if the principles are properly adhered to or calculated and only if the resultant solution is manufactured, operated, and maintained exactly to the standard required by the design.

Similarly, those considered least effective—such as administration controls and personal protective equipment—can be enhanced in their efficacy by the introduction of further control measures such as training, supervision, authorization, empowerment, and culture. These are not failure proof, obviously, but correctly designed and maintained administrative systems can be just as effective (if not more so) than poorly designed, maintained, and inspected engineering solutions. The most important thing is that the *blend* of control measures is correct for the undertaking to which they apply; as is their proper maintenance, management, and revision when necessary. Getting the correct blend of controls is likely only possible when there is a full and proper understanding of the proposed control measures and any flaws inherent in them.

It may be seen, then, that even where an assessment of risk has been made and control measures have been put into place to mitigate the risks identified, there is still the opportunity for peripheral risks to create fault lines in those controls. A straight-forward risk assessment requires only a straightforward solution: a sort of call and response if you will. What is required is something that looks beyond any suggested control measures to the supplemental risks inherent in those measures. To see how an operational risk assessment might be enhanced by a more in-depth assessment of inherent risks, let us consider a large hydraulic power press as may be commonly found in engineering manufacturing

Tables 5.2–5.5 compare some illustrations of what we may normally see in a risk assessment for this type of machine with what we should ideally be thinking about in extended terms of the risks posed. The reader may well be able to imagine some others. It should be evident that this depth of assessing risk provides an additional layer, or layers, of safety, over and above what a normal operational assessment might provide. And it begins with the criteria, operational limitations, and safety factors originally set in the design and manufacture of the machine, structure, process, or whatever. The issue is that the design safety case is rarely disseminated to the end user or operator. The safety practitioner rarely has insight into these design elements.

Table 5.2 Safety and risk analysis—electrical supply

Standard risk assessment	Safety and risk analysis
Fixed supply is fitted by qualified electrician	Is there sufficient capacity in the supply to run the machine?
Cable management keeps cables away from impact, etc.	What regulatory inspections are required for the machine, cabling, and supply?
	Can the cable management cope with the voltages/amperage/heat in the cables?

Table 5.3 Safety and risk analysis—guarding

Standard risk assessment	Safety and risk analysis
Fixed guarding is in place around the sides and rear of the machine	Does any guarding need to be removed to perform live testing or maintenance on the machine?
An automatic guard is provided on the open side of the machine	What provisions are there to secure the guarding against wear, vibration, etc.?
	Are there failsafes built into the automated guard
	Are there any single points of failure on the automated guard, such as cotter pins, etc.?

Table 5.4 Safety and risk analysis—operation

Standard risk assessment	Safety and risk analysis
Operatives have received suitable training	Is training updated, monitored, and refreshed as required
Only trained operatives have a key to start the machine	Is the person who keeps the training register updated covered in the event of absence
	How is key access controlled
	Is any form of operation or training reliant on one individual, and what cover is provided in the event of their absence

Table 5.5 Safety and risk analysis—maintenance

Standard risk assessment	Safety and risk analysis
The machine is maintained to manufacturer's specification	Is maintenance reviewed and refreshed in line with the use and age of the machine
Maintenance engineers use "lock out tag out" system	How is the lock out system controlled if the engineer has to run the machine during testing

This is where the "golden thread" of information is important, but its proposed use in a narrow segment of just one industry—higher-risk buildings in the construction sector—belies its significance in *all* forms of construction, engineering, and manufacturing.

The health and safety file has only had a real application in "traditional" construction (and even then, with regrettable scarcity), and the golden thread has been proposed for a small area of construction, i.e., that concerned with higher risk buildings. The safety case has been around for a very long time and has been used for all manner of large scale and high-risk projects. Its validity lies predominately in its structure of claim-evidence-argument, which forces the author of the safety case to justify their definition of "safe". This is in contrast to being purely a list of historical events and decisions without causal reasoning.

5.6 Discovering risk

In any given undertaking, there could be many forms of assessment of risk pertinent to various functions of the organization. These various assessments are most likely completed by different individuals with differing levels of expertise and disparate and seemingly incompatible qualifications. The outputs of these assessments can also vary in size, complexity, and terminology and will invariably be used by different facets of the organization to make estimations or forecasts, to establish exposure to various risks, or to inculcate safety procedures or processes. A list of possible types of assessments, as well as those who may be responsible for them and those who may benefit from them is given in Table 5.6.

Table 5.6 Types of assessment

Type of assessment	Who might originate it	Who might benefit from it
Safety case	Designers	Designers, operators
Risk assessment	Operators	Operators
Risk register	Management	Operators
Quality assessment	QA, management	Operators
Business risk	Executive board	Business

It is worth noting that in *Acceptable Risk*, the authors believe there is no such thing as an objective evaluation of risk (Fischhoff, Lichtenstein, Slovic, Derby, & Keeney, 1981). Rather, an assessment of risk at best represents "the perceptions of the most knowledgeable technical experts". The crucial element here is that, whilst anyone may be able to identify an obvious risk—such as a poorly stacked pile of bricks, for example—their individual skill sets and experiences will cause them to treat the risk with varying levels of concern. This is especially the case when there are other readily identifiable risks in the vicinity to which the individual may attach more importance. Similarly, a risk connected to safety or finances may be obvious to many, but the consequences beyond the obvious—such as prosecution under some regulation or other—may only be apparent to those specifically trained in that particular field. The proper, effective assessment of risk should therefore involve as many stakeholders who may be affected by that risk and its consequences as necessary.

The common factor in all of examples in Table 5.6 is, of course, risk; the quantum of likelihood and severity of any particular outcome. Risk is invariably considered in the negative sense—the risk (or likelihood) of something going wrong. However, in risk management terms, risk can have positive connotations too. The risk from purchasing a lottery ticket is, for example, that you might win. There is also a risk, of course, that you might not; but without purchasing a ticket, there is no risk of either event occurring. In overall risk management terms though, the actual purchasing of the ticket is only one small part of the total exposure to risk. Leaving the house, driving the car to the shop, walking to the door, interacting with other people, the weather conditions, the time of day, the number of other people buying tickets that day—all of these events have an element of risk attached to them. In the grand scheme of our daily lives, however, we generally pay scant regard to them. But if one were to sit and write down each individual element of the purchasing process, calculating the risk factors for each, you would find that the actual odds of winning the lottery would be affected, albeit perhaps only very slightly.

In health and safety terms, there is a process called job safety analysis whereby any work process is broken down into individual component parts. Each of these component parts is then assessed for risk. When completed by health and safety practitioners, this would be following the IDERR methodology: *identify* the hazard; *decide* who is at risk of harm and how; *evaluate* the risk and ways to control it; *record* and *review* the assessment. By doing this, a picture can be built up of the entire work process and the salient levels of risk involved at each point in it. In a workplace risk assessment, this should encourage the relevant resources to be directed where there is higher risk. That is to say, where the likelihood and severity—or amount of harm—are greatest. Let us look at the process of replacing the wheel on a car as an example (see Table 5.7).

Table 5.7 Steps for replacing a wheel

Step	Action
1	Apply handbrake
2	Chock wheels
3	Remove spare wheel, wheel brace, and jack
4	Position jack
5	Loosen wheel nuts
6	Raise jack
And so on	

5.7 Risk assessment

We can now add the risk assessment element of the process. For clarity, we have dispensed with identifying the hazard (the car, gravity, physical force, etc.) so that we can concentrate on the risks at each stage (see Table 5.8).

Table 5.8 Risks associated with the steps

Step	Action	Risk
1	Apply handbrake	Uneven ground may cause car to roll
2	Chock wheels	Chocking the wrong wheel may cause chock to fail
3	Remove spare wheel, wheel brace, and jack	Strain or injury to operative lifting out items
4	Position jack	Injury to operative positioning jack. Operative laying in road whilst performing task
5	Loosen wheel nuts	Strain or injury to operative whilst undoing wheel nuts
6	Raise jack	Crush injury to operative if car rolls off of jack or jack collapses
And so on		

In a workplace risk assessment, we would now look at control measures for these risks (see Table 5.9). These measures should follow the hierarchy of control of: *elimination, substitution, engineering controls, administrative controls, personal protective equipment*. This is sometimes referred to with the abbreviation *ESEAP*, and is a simplified version of the general principles of prevention (Directive 89/391/EEC, 2021).

Table 5.9 Control measures associated with the risks

Step	Action	Risk	Control measure
1	Apply handbrake	Uneven ground may cause car to roll	Only complete task on firm, level ground
2	Chock wheels	Chocking the wrong wheel may cause chock to fail	Chock both wheels on opposite side to wheel being replaced
3	Remove spare wheel, wheel brace, and jack	Strain or injury to operative lifting out items	Engage correct lifting techniques. Use two people to lift spare wheel from boot
4	Position jack	Injury to operative positioning jack. Operative laying in road whilst performing task	Close off live traffic lane adjacent to car. Have an assistant act as look-out. Wear high-visibility clothing. Wear gloves with tear and impact resistance
5	Loosen wheel nuts	Strain or injury to operative whilst undoing wheel nuts	Ensure tools are in good condition before use. Use correct pulling technique
6	Raise jack	Crush injury to operative if car rolls off of jack or jack collapses	Ensure jack is in good condition prior to use. Stand clear of car whilst raising the jack. Wear safety boots and gloves
And so on			

You may see from the above example the sort of risk assessment with which we may all be familiar in the workplace. It is certainly the sort of thing that anyone with relevant health and safety training might arrive at. However, it makes several assumptions for which there is no assessment of risk, primarily because there is no information with which to assess the risk posed. We do not know, for example, where the task is taking place, and we cannot, therefore, be in control of assessing the surroundings. We do not know the precise location of the spare wheel or the tools required to perform the task. The type of jack is also unknown, as is the level of competence of the operator. This may seem all rather trite in comparison to the task itself, and some might say that overloading the assessment of risk from a task that many of us have had to perform in the course of general car ownership over the years is overbearing at best. But let us extrapolate this example to, say, Piper Alpha. There, a number of trained operatives performed a task for which they were more than competent and had performed many times before. Neither shift involved in the events that led to the disastrous fire was engaged in anything that was out of scope or novel to their defined procedures. Admittedly, there are few industries that pose as many risks as the deep-water extraction of volatile resources, but the fact remains that some seemingly innocuous undertakings still carry the *potential* risk of causing serious harm. Lifting a vehicle weighing perhaps two tonnes into a precarious position whilst working alongside it is most certainly one of them.

And this is the important point to make. Writing one email whilst crouched over a laptop on the settee is unlikely to result in any long-term or life-threatening issue. Performing the same task all day, every day, for five years will very likely result in some musculoskeletal condition that may prove life changing, possibly even life limiting. Similarly, each task undertaken prior to the fire on Piper Alpha had been assessed and, it could be argued, suitably controlled. But the *concatenation* of events, or tasks, each with one or more potential failure points, can occur in a way such that those points align, hence allowing a direct path to a harmful outcome. It could be argued that in the case of Piper Alpha, as in the case of many other similar disastrous occurrences, that the major flaw was in the interstice between the main events: in this case, the information transfer between shift changes. But whether a direct path to a harmful outcome is the result of some interstitial failure or shortcoming; or whether it is a result of one of the central tasks not being conducted properly (for whatever reason), the key to identifying potential safety failures is in understanding not just the task but the peripheral factors surrounding it. Let us call them the *environs* of the task. Let's return to our example to see, in Table 5.10, how these environs for each component of the task might look.

Table 5.10 Peripheral factors (environs) associated with each step

Step	Action	Risk	Control measure	Environ to the risk
1	Apply handbrake	Uneven ground may cause car to roll	Only complete task on firm, level ground	We have no control over the ground type. We cannot affirm if the car or vehicle are therefore suitable
2	Chock wheels	Chocking the wrong wheel may cause chock to fail	Chock both wheels on opposite side to wheel being replaced	Are the chocks of a suitable size, shape, and material? Can they be placed and removed safely (handles)?
3	Remove spare wheel, wheel brace, and jack	Strain or injury to operative lifting out items	Engage correct lifting techniques. Use two people to lift spare wheel from boot	Is the wheel available without having to remove the contents of the boot? Is there a lifting aid to withdraw the wheel?
4	Position jack	Injury to operative positioning jack. Operative laying in road whilst performing task	Close off live traffic lane adjacent to car. Have an assistant act as look-out. Wear high-visibility clothing. Wear gloves with tear and impact resistance	Is the jack of a suitable type to lift the car? Can it be placed and withdrawn safely (handles)? Can the lifting/lowering mechanism be accessed safely?

(Continued)

Table 5.10 Peripheral factors (environs) associated with each step *(Continued)*

Step	Action	Risk	Control measure	Environ to the risk
5	Loosen wheel nuts	Strain or injury to operative whilst undoing wheel nuts	Ensure tools are in good condition before use. Use correct pulling technique	Is the wheel brace able to deliver sufficient torque? Is the handle extendable, rubber-coated?
6	Raise jack	Crush injury to operative if car rolls off of jack or jack collapses	Ensure jack is in good condition prior to use. Stand clear of car whilst raising the jack. Wear safety boots and gloves	Is the car's jacking point suitably strong? Is it properly indicated? Can it resist moments of force should the mass of the car shift?
And so on				

It can be seen from the deeper issues connected to each component of the task that any one of them could, should they fail (that is, not be to the required standard, type, material, etc.), present a clear pathway for a harmful outcome to occur. This outcome may not necessarily be the worst-case event of a crush injury from the jack collapsing—it could be just some badly rapped knuckles—but any harmful outcome is contradictory to the aim of assessing risk and promoting safety. (Although this does not take account of any level of acceptability of risk, the flip side of the phrase *as low as reasonably practicable*. We may, for example, not mind the odd scraped knuckle, in which case our control measures will be relevant to reducing risk to this level of potential outcome. This is the concept of risk tolerance and will be discussed later in chapter 7.17.)

It may be considered a little extreme, possibly indulgent, to explore the "back story" of risks in this way, particularly for such an apparently straightforward task as changing the wheel on a car. Besides, one might argue, how many wheel changing operations result in someone getting hurt? This is to ignore several important points that we are making here. Firstly, it is impossible to say how many times simple, everyday tasks result in some form of harm. A cut finger usually results in a person applying a plaster; something more serious may end up in a visit to hospital, and that would only be recorded if the injury was sustained whilst at work. Secondly, any harmful outcome is contradictory to the very essence of "safety". Certainly, we are trying to ensure that the risk of a harmful outcome is kept *as low as reasonably practicable*, meaning that although a harmful outcome is still *possible*, it is not the *expectation* of the control measures we put in place. Thirdly, although our example is fairly benign, it is reflective of many similar, straightforward tasks that have resulted in regrettably serious harm. It should also be considered that the reason that the changing of a spare wheel on a car is not more widely known as a serious threat to a person's safety is that the engineers and manufacturers involved have conducted the sort of safety analysis that we are promoting here. Finally, we should consider that it is often when performing seemingly benign tasks in succession that a failure of control measures can occur, particularly when one includes ancillary issues such as complacency, poor culture, ergonomics, stressors, and so forth.

These ancillaries to the main risks are what invariably cause the direct path from risk to harmful outcome: the "short circuiting" of the control measures if you will. They are perhaps no more apparent in the causation of harmful events than in air transportation. Far more complex than our example of replacing a spare wheel, commercial air travel is among the safest forms of travel. But being tens of thousands of feet up in the air, travelling at hundreds of miles per hour in a metal tube is not a tenable position for human beings, and when things go wrong they can do so disastrously. Modern airliners are fitted with a plethora of safety features, and the training of their pilots is highly intensive, including practising for emergencies in flight simulators. The use of simulators started before the First World War and allows pilots to be not only trained in normal flight but also tested against abnormal and exceptional situations, all without the inconvenience of gravity getting in the way. But despite the levels of training, the implementation of safety protocols, the existence of numerous safety devices, and the standards of maintenance on the ground, there are still air disasters from which we can learn lessons. This is part of the inescapable feedback loop that must exist in all assessments of risk where, despite our best efforts, a harmful outcome is realized, and we must return to, and review, our control measures.

In aviation terms, an archetypal example of flaws aligning in the accepted control measures for flight safety was the crash of flight AF4590 in July 2000. The aircraft, a Concorde operated by Air France, had had an enviable safety record for 31 years prior to the crash. Immediately following the incident, all Concordes were grounded and following a thorough investigation extensive modifications were made to the fleets operated by both Air France and British Airways. It was originally discovered that the aircraft's failure was initially caused by it running over a piece of debris, which had fallen from a previous flight as the aircraft was about to take off. The debris, a strip of metal around 400 mm long, burst one of the Concorde's tyres causing tyre fragments to impact with a rear fuel tank, rupturing it. The resultant spillage of fuel was ignited by the jet plume. Stronger tyre compounds, burst-proof fuel tank linings, and strengthened wiring looms were all retro-fitted in order to prevent this type of event reoccurring. Subsequent investigations, however, discovered several failings that were a precursor to the tyre's impact. An axle bushing had not been replaced during a recent regular service of the aircraft causing one of the sets of landing gear to run out of true. The aircraft was in excess of its maximum structural capacity due to having been overloaded with baggage. This was exacerbated by the fact that the pilot chose to take off with an 8-knot tailwind which translated into the aircraft being six tonnes overweight as well as having too high a percentage of its load aft of its centre of gravity. The aircraft also took off 11 knots below its minimum take-off speed and, due to the fire, had one of its engines apparently shut down by the flight engineer at just 25 feet above the ground (Rose, 2012).

All these issues—maintenance, loading manifests, pre-flight checks, and calculations—are, and have been for many years, tightly controlled in aviation. This is central to the fact that these sorts of events happen, thankfully, very rarely in comparison to the many millions of flight miles racked up by commercial airliners. The absence of virtually any one of these flaws could have enabled the aircraft to be returned to the ground safely. Indeed, it is considered that if just the axle bushing had been replaced as it should have been, the aircraft would have taken off before encountering the debris on the runway (being out of true, the axle was causing additional tyre drag which meant the aircraft needed more runway to achieve take-off). The issue of the flight engineer apparently shutting down one of the engines is an example of where technology might have prevented such a situation occurring. But technology can be a double-edged sword in terms of safety for the reasons of training and complacency. This was a situation that, staying with aviation, troubled Boeing on two separate but connected sets of circumstances.

The Boeing 777 was designed to have some of the most advanced safety features of any commercial airliner at its launch, including an advanced multi-function autopilot system. A particular set of circumstances nonetheless revealed a flaw in the layers of control measures this offered. Flight OZ214 from Seoul was coming into land at San Francisco International Airport in July 2016. The airport's vertical guidance system, or guide slope, was not in operation at the time, and this had been communicated to all flights arriving at the airport. The guide slope device allows pilots to accurately gauge their approach angle relative to the runway. It is one of several systems pilots use on their approach and many hundreds of aircraft had landed successfully during the time the system at San Francisco had been out of service. On approach, the pilot of flight OZ214 realized the aircraft was too high relative to the correct approach path and disengaged the autopilot system in order to lose some speed and, therefore, height in order to correct this approach angle. One of the several functions of the autopilot system is a speed monitor—akin to a cruise control in a car—which automatically applies power to the engines to maintain a speed set by the pilot. Although this approach speed had been set by the pilot, he was not aware that this function of the autopilot was disengaged when he switched off the autopilot system. Unaware that his aircraft was losing more speed than he planned, as well as height, it was not until a few seconds prior to landing that the crew realized they had lost too much of both, and the aircraft struck the retaining wall of the runway, which at this particular airport extends out into a body of water (NTSB/AAR-14/01, 2013).

Despite the intentions of such an advanced system, there had not been sufficient explanation given by the manufacturer of its limitations, albeit in a very precise set of circumstances. This situation regrettably happened again only a few years later with the launch of the Boeing 737 Max aircraft which, once more, promised new levels of safety and advanced avionics. The introduction of the Manoeuvring Augmentation Characteristics System promised high levels of safety by preventing pilot errors that could lead to a loss of control. Unfortunately, the information on the system given to pilots was not fully explicit on its operation, and in very particular circumstances, it could countermand perfectly valid inputs from the pilot. This was in part due to there only being one sensor for a particular piece of data: that of the aircraft's angle of attack. If this one sensor were faulty or compromised in any way, the aircraft's computer had no way of reconciling this information and could therefore adjust the aircraft's pitch erroneously. This is what happened to two flights of the new 737 MAX-8 aircraft in October 2018 and March 2019, resulting in the tragic loss of a total of 356 people (PMC7351545, 2020).

The exponential increase in reliance on technology over the last few decades has brought many benefits. Driverless cars; autopilot systems; train protection systems—all of these have impacted positively on safety in transportation, as have robots, scanning systems, light guards, and a host of other technological advances in industry. One issue that comes with a technical advancement, *any* technical advancement, is that it becomes de facto; the paradigm. Where the resultant and promised level of safety that the advancement brings is accepted as the norm is where human behaviour (in the form of availability heuristics, confirmation bias, and good old, traditional complacency) can detract from any improvements in safety. The early application of train safety, for example—the introduction of block operation and men with stop-watches—led to improvements in safe train operation, which were subsequently eroded as train companies reduced the lead time between trains to increase passenger numbers.

Fundamental safety improvements, that is to say, those which rely on solid engineering principles rather than advanced electro-mechanics or electronics, tend to have better, more long-lasting positive effects on safety. In road transport, such things as seatbelts, speed limits, roundabouts, and crash helmets have made marked improvements to road safety and the reduction of collision casualty rates (Road Safety Annual Report, 2015). There is no denying that advanced avionics have made air transport safer, but this must be considered alongside the fact that airline pilots are amongst the most highly trained of any workers. And even here, we see that some human behaviour types and characteristics can still affect these advanced systems. The Aviation Safety Whistleblower Report (US Senate, 2021) in December 2021 by the US Senate Committee on Commerce, Science, and Transportation noted the comments of a former Boeing engineer about the complexity of human factors in connection with advanced technical systems:

> That complexity stems from the nature of the pilot's task of integrating an understanding of automated functions and the aerodynamic state of the airplane; a pilot uses everything from indicators, control force feel, "seat of the pants," prior experience, and training to develop appropriate responses to whatever situation arises.
>
> (Curtis Ewbank)

This indicates the need for *any mechanism* integrated into a working practice for the benefit of safety to be specifically designed around the inputs of the individual—skills, knowledge, aptitude, training, and experience (*SKATE*)—in *combination with* the anticipated or required safety-related outcome. In their book on organizational safety, Dr Waddah Al Hashmi and Bob Arnold refer eloquently to "a strategy of redundancy beyond technology" (Ghanem Al Hashmi & Arnold, 2021). In terms of safety, this is the verification that the means support the ends; that the technology is—and remains—fit for the purpose of improving safety. This verification can be stretched by some technologies which are well-intentioned, well sold, and well implemented and become a de facto component within a safety system. This may be by way of a piece of software, a reporting system, a monitoring device, or some technical apparatus. By sheer dint of the duties placed on the designers of any of these things designed to promote safety, they should be properly and adequately designed to fulfil their role "without risk" (HASAWA, 1974). But what if the device or software has been created by the manufacturer for purely profitable purposes? What if it comes from a country that does not have such a robust health and safety legislative culture as a country like the UK, Germany, or the USA, for example? And what if the customer was misinformed about or misunderstood its application or benefits?

We discussed earlier the difficulties encountered when monitoring safety in an organization; the possibility of using a reporting methodology to indicate that all is well at the visible level when, in truth, there are underlying errors and omissions that go unnoticed or unreported. It is quite understandable that an organization can come to rely on technically advanced systems as a panacea: and if the culture of the organization is not mature enough to examine the details more closely, these systems become endemic to that culture. We can see the results in general work risk assessments, where these systems, devices, and programmes are integral to the control measures. We have, contrastingly, also explained how it is beneficial to examine in detail the control measures that are, or will be put, in place by identifying the *control measures within*—what we called the *environs*. And despite the danger of repeating ourselves, this identification process should be cognizant of, and appropriate to, the level of risk involved. Some examples might include those identified in Tables 5.11–5.13.

Table 5.11 Risks associated with an office

Copier paper to an office		
Type of paper	Is it appropriate for all types of printing demands?	How is the paper tested for use?
		Are manufacturer's recommendations taken into account?
	How is it packaged?	Is the paper available in the quantities required?
		Who will be required to handle the paper, and is appropriate for manual handling?
Storage of paper	How is the paper stored and delivered to site?	Is sufficient space for the size of delivery vehicle?
		Is the paper stored appropriately with regard to humidity, UV, and temperature?
		Are there requirements to acclimatize the paper prior to use?
		Are there additional hazards during storage: liquids, sources of heat, etc.?

Table 5.12 Risks associated with climbing

Climbing ropes for mountain rescue personnel		
Strength of rope	Is it appropriate for all sizes of personnel?	Are the ropes coded for strength?
		Is it viable to have varying strengths for different personnel or just the strongest rope available?
		How is rope tested, and to what extent?
		Where is the rope tested, and what is their competency?
		What is the protocol for quarantining defective rope?
Storage of rope	How is the rope stored and delivered to site?	Is storage to manufacturer's recommendations?
		Is it checked in storage, how often, and by whom?
		What is the protocol for quarantining defective rope?
		Are there additional hazards during storage or transport: humidity, fumes, acids/alkalis, UV light, abrasive surfaces, etc.?
Use of rope on site	What are the rules and practices for rope work?	Are only certain personnel trained to use the rope, and if so, how are they monitored?
		What pre-checks are carried out, and how are these monitored?
		Who has the authority to abandon use, and how is this recorded?
		Who is responsible for transportation, setup, ground conditions, etc.?
		How is competency gauged, and how is this recorded?

Table 5.13 Risks associated with food preparation

Manufacture and supply of chilled food products to the public		
Ingredients	Are the ingredients traceable?	How are suppliers selected?
		What initial and ongoing validation of suppliers takes place, to what standard, and how is this recorded?
		What checks are performed on delivery to validate the ingredients?
		What is the protocol for accepting temperature-controlled deliveries: standing time, acceptable limits, etc.?
		Who is responsible for rejection of ingredients, and how is this recorded?
Storage of ingredients	How are the ingredients stored on site?	Is storage to supplier's recommendations?
		How are temperature-controlled facilities checked, by what method, and how is this recorded?
		What are the acceptable limits of storage temperature?
		Who is responsible for storage failures, quarantining of products, and rejection of stock, and how is this recorded?

(Continued)

Table 5.13 Risks associated with food preparation *(Continued)*

Manufacture and supply of chilled food products to the public

Manufacture	What is the manufacturing process, and how is it controlled?	How is machinery cleaned, maintained, and repaired?
		Who is responsible for cleaning, maintaining, and repairing, and how is their competency gauged?
		What are the methods of checking temperature during production, and how is this validated?
		Who is responsible for rejecting batch runs due to process failures, and how is this recorded?
		How are products tracked after shipment?
		What is the protocol for recalls in the event of production issues?
		What is the protocol for quarantining and disposing of failed production batches?

It can be seen from these three examples that the level of scrutiny of the layers of controls has increased with the relevant level of risk from each undertaking. The delivery and use of copier paper to an office may cause a musculoskeletal strain or injury if the package is not handled correctly for its weight, or it may pose a fire hazard should it be stored inappropriately. It is unlikely though to cause in itself any major risk to safety. A climbing rope being used to belay a rescue worker down a cliff could lead quite easily to a fatal event: possibly even two fatalities if the rope fails as the rescuer and injured party are being hoisted back up the cliff. The production of chilled food items is fraught with a multitude of instances where the ingredients or finished product could become contaminated. Not only with foreign objects from the processing or production stage but also from any of the dangerous bacteria that can flourish if the proper storage and handling conditions are not maintained. This could affect hundreds or perhaps even thousands of people, not to mention any reputational damage to the organization as well.

The examination of detail behind any control measure is possibly something that many of us do as a matter of course in any event. As with designing something safely, it is not unreasonable to presume that any competent, professional designer will inherently assess their design for risks and mitigate them through their knowledge, experience, and training. What is a common failing, however, is the *formal recording* of this mitigation. And from this, the direct dissemination of this mitigation to the manufacturer, constructor, installer, owner, and user of that design output. This is the successive, progressive, updated, and recorded sequence of information that can be used to identify issues more readily; implement maintenance plans more adroitly; inform users and operators more accurately, and, identify causal events in the case of accidents and failures more quickly. It supports the term "golden thread" which for some has become so important. And yet it stems from systems and processes that already exist and have done for decades.

6 Building the case for safety

6.1 Identifying the hazards

The method of demonstrating safety in any given function—be it operational, organizational, financial, construction, and so forth—varies in phraseology, form, and output. The
core reason for the demonstration, however, is the same: to identify the hazards in the
function; assess the risks posed by those hazards; and provide suitable control measures
to mitigate those risks. No one methodology might be considered better than any other
because they all have their merits depending on the specific function that one is assessing
and the type of documentary output that is expected. What might be realized though
is that these various methodologies are rarely seen as cohesive, or even supportive of
one another. Different professions have evolved their own methods of assessing quintessentially the same information but in different ways using different modifiers and nomen-
clature. This is unhelpful because the information captured in one method of assessment
might be highly valuable to another method if, firstly, each method was aware of the
presence of the other, and secondly, the two communicated with one another.

One area of dissimilarity is in the language used to mitigate risks. Control measures; arrangements; resources; mitigation: all of these refer to ways that risks can be
controlled, reduced, or fended off. There are hierarchies of control, principles of prevention, and risk management techniques that are all, ostensibly, the same thing but
with different substructures and processes. They are the same because they all require
accurate assessment of the risks and proper introduction of the control measures they
advocate. Another telling similarity is that whatever control measures are advocated
by a properly conducted risk assessment (or risk analysis, or risk profiling, if you prefer), they will almost always have risks inherent within them. Flaws, cracks, or holes
in the control measures that might, under the right conditions, allow the risk they are
meant to be controlling to pass through and become realized as a harmful outcome.
That is to say, the controls fail and the consequence of the risk becomes a reality.

This model is much used and possibly a little clichéd, but it still perfectly represents the
issues with any type of control measure. Reason's analysis, of course, was with regard
to accident causation, and how this could be controlled. The point of this book is to prevent accidents happening in the first place, and we should reiterate that in this context
"accident" means an unplanned or uncontrolled event that leads to a harmful outcome.
That harm might be an injury, a loss of production, a financial cost, or some other loss or

DOI: 10.1201/9781003296928-6

harm that is undesired or unplanned. (Cross reference this with scheduled maintenance, for example. A machine being maintained will not be producing anything, but this is not an accident because it is planned and accounted for.) What is also to be understood is that any accident in these terms can have causational effects on other accidents occurring: that is to say, other losses or harmful events. The Concorde crash of 2000, for example, led immediately to the tragic loss of life of all those on board. It further led to loss of life on the ground (the aircraft struck a hotel) and to the destruction of property. Subsequently, it led to a loss of income as the whole fleet was grounded temporarily, and then later still to huge unplanned expense in making modifications to the fuel tanks and tyres. Eventually, the entire fleet was retired early, leading to a further loss of income. One might also add the reputational harm done to one of the most advanced and iconic aircraft ever built. It also had even further consequences in that travellers who had come to rely on the supersonic flight times across the Atlantic were now confined to standard subsonic flight times, impacting on their ability to conduct business in the manner they had become accustomed. The total cost of the impact from this one accident is unlikely to ever be known but is undoubtedly extremely considerable.

Impact upon impact; consequence following consequence. This is the real cost of accidents, in terms of all losses, which is rarely considered by the general risk assessments and analyses that we conduct on a daily basis. We refer to these as the *environs* of control measures: the factors that affect each control in varying ways and degrees, just as each measure affects the risk itself. And as with each control measure, the environs of each are also affected by flaws that can, in the right circumstances, cause the control to fail. The factors affecting control measures are often attributable to such things as documents and documentary systems, organizational issues, and human factors. Even something as relatively benign as the periodicity and efficacy of a review process can have drastic consequences. Consider a machine at which an operator works which has an opening that could allow the operator to place a hand inside during the machine's operation. A risk assessment would, rightly, identify this and perhaps call for a fixed guard to protect the worker. But one day, the machine is required to process some material that does not fit through the machine with the guard in place. The material has been delayed in arriving at the factory and the worker is under pressure to finish the task of processing it. The risk assessments are reviewed perfunctorily every six months, but the next review is three months away and the safety officer is currently on holiday. It only takes the undoing of four screws to remove the guard—and the processing of the material should not take too long anyway.

It is situations like this, where everything appears to be as it should, that the flaws in the environs of the control measures become evident, and harm occurs. Many organizations find themselves in a situation of harm or loss which they find genuinely shocking, certain as they were of the systems they had in place and the efficacy of them. This is not to say that we can avoid every possible harmful event, nor should we be attempting to do so at any cost. The importance of understanding *consequence* is, as we have discussed, the ability to prioritize the efforts in controlling risk. Whether this is via a quantitative method, such as probability analysis, or through a qualitative one based on experience and subjectiveness is entirely dependent on the risk in question. The failure of a pressure valve in a chemical process is likely to be subject to such analysis of its mean time between failure rate (MTBF), which will undoubtedly affect the maintenance regime and replacement strategy. Investment in a new market by a growing business will perhaps be affected by the directors' experiences beforehand and their collective appetite for risk versus growth. Both situations are subject to the proper identification of the risks—and probable consequences—as well as the effective control measures that should be in place. But both are still likely to fail for any manner of reasons which may or may not have been identified, or found to be too costly to prevent. Notwithstanding this, standing blinking in the harsh light where there used to be a factory, wondering what on earth that loud noise was, is not a situation that any organization wants to be in, and certainly not without very good reason.

The influence of human factors on safety has been a subject of academic study and discussion for decades, starting primarily with the concept of how human error was a primary causation of accidents. And although this book is not about accident investigation but *prevention* through the proper application of safety, it is interesting to see the progression made from these early analyses: Rasmussen's skill-rule-knowledge-based model; Hale and Hale's behaviour model; Hale and Glendon's model of behaviour when faced with danger; and Reason's model of system safety. This latter model is reimagined in Figure 6.1. Of note are the "defence barriers" on the right-hand side which demonstrate the effectiveness, or otherwise, of additional control measures applied after the inherent controls of management, training, organization, and culture.

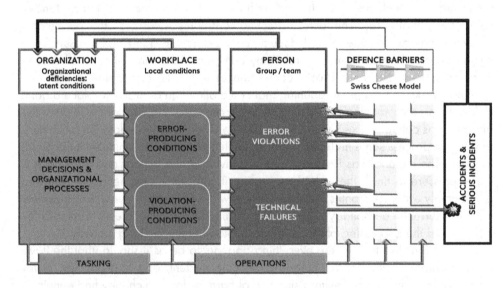

Figure 6.1 Reason's model of system safety.

It is this model that resolves the influence that external factors have on human behaviour, perhaps most succinctly, in that even as we have discussed, some evidently dangerous—or *unsafe*—situations can be dealt with safely depending on many and various input factors surrounding the situation. These factors include such matters as training; support; determination; character; equipment. In resilience engineering, safety is considered more than just an absence of failure, and the functional resonance analysis method (FRAM) developed by Erik Hollnagel (Hollnagel, 2012) is a method of analyzing why things could go wrong (or right) depending on a number of influencing factors such as preconditions, resources, controls, and time. Again, it is about understanding the background to the control measures applied, with reference not only to what is applied but why they are applied, by whom, and how they are responded to. In our example of operating a machine where the guard needs to be removed in order to process an oversized piece of material, the operator stands a greater chance of returning home with all their fingers if they feel that they are working in an environment where they can call into question a safety-related matter quickly, and with the right person, rather than just having to "get on with it".

What is clear from any of the methods or models for affecting or influencing behaviour is that understanding the baseline from which one is working is key. In the same way as accurately identifying the risks in risk assessments, understanding the process in a job safety analysis, or collating the evidence for a safety case. We must also ensure that the information we glean during this process is reusable in a variety of other formats. That is to say, it can be used by the engineering department to substantiate their safety case, by the safety department to be used in writing risk assessments, or by the finance department to gauge the ability and appetite to venture into a new market. Until now, these disparate groups have used their own methodologies and phraseologies to establish their own criteria for assessing risk. But the central tenet is that safety in *anything* is about reducing the risk of a harmful consequence occurring to *as low as reasonably practicable*. And the safety inherent in any organization is based on factors that start with the owner or managing director or CEO and continue through every level, component, and function of that organization.

6.2 Adding the "argument"

In general terms of risk management, be it in safety, financial, or operational areas, for example, we support our belief that things are "safe" (for a given value of "safety" or so far as reasonably practicable) with an assessment of risk. The style and content of this assessment may vary depending on the area being assessed and the individual doing the assessment, but, in the main, the core elements of them are the same: identify the risk(s); understand the consequences; recommend suitable control measures. Our determination to make something "safe" is therefore our claim, and the evidence we provide to support that claim is the assessment of risk. With an engineering safety case, however, we must take another step and prove that the evidence is what we say it is and that it actually supports the claim.

In Figure 6.2, we see how a claim and its evidence may fit snugly together, but there is a hole, a gap in the pieces which, if pulled at, would cause the claim and evidence to fall apart, just as these two pieces of a jigsaw puzzle would. We may have written a wonderfully complete and exhaustive risk assessment for some task, but if nobody

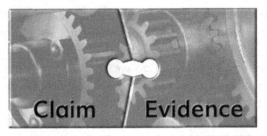

Figure 6.2 The fragility of evidence supporting a claim without the relevant argument.

reads it or complies with it, then it is not supporting our claim (that is to say, our belief) that the task is safe. The reasons that nobody reads the risk assessment will depend on any number of the factors surrounding the control measures (their environs) that the assessment has recommended: how people are trained or supervised, for example, their level of morale, or their ability to access the document itself. Note that we have not suggested that the document has not been read because the workforce is lazy, or belligerent. We do not immediately presume that it is the workers who are the problem. And even if they were lazy or belligerent, would that not be as a direct result of the conditions in which they work: the management, culture, welfare, and so on?

Implementing the arguments to support the evidence of a claim starts with understanding what those environs are—what influences the controls we have in place and what, if anything, we need to do about them. Some influences might be less apparent than others, and some might be things over which we have little or, perhaps, no control. This is completely acceptable because we cannot expect to have full control over every aspect of risk, but understanding *where* the potential for failure of our control measures might lay helps to illustrate any further actions we need to take. In essence, if we know that there is a weakness in one control measure, we can address this by increasing the resilience of another control measure to compensate. These would inevitably be based on the standard risk control strategies of:

- Avoidance—abandoning the process altogether
- Reduction—introducing further measures to control the influences identified
- Transfer—moving the ownership of risk, for example via insurance
- Retention—accepting the level of risk

We may assert, for example, that operators are safe to use a particular machine because they have been trained to do so and that the risk assessment identifies this and states the type and level of training required. This is our claim and our evidence. The argument in support of this might be that we can demonstrate that supervisors are all time-served with the organization and therefore have an innate knowledge of its ambitions and standards. We may demonstrate that the training is provided to supervisors first to make sure it is fit for purpose and that records of training are held by the human resources department in an electronic form that automatically generates reminders for refresher training. We could point to authorization notices displayed near each machine noting who is authorized to operate which machine. We could highlight the safety tours undertaken by the production manager each month and the results of their findings, as well as the minutes from production meetings that discuss training, productivity, safety, and feedback from the operators. All of this would constitute the argument in support of the claim and evidence and, of course, could be used as the argument in support of many other tasks in the organization.

The argument provides the missing piece that binds our claim (the belief that we are "safe") with the evidence (the safety-related documents that we generate), just as we see in Figure 6.3. The crucial part of creating the argument is understanding *what* can influence the evidence.

Figure 6.3 The relevant argument that supports a claim's evidence.

6.3 Key factors

There are several core factors that are key to understanding the environs to any existing or planned control measures. We have grouped these into four sections, each with a number of subsections relevant to the section's heading. As with any type of system, there are two caveats: firstly, the system should only be as complicated and complete as the organization it is describing, and secondly, it will only ever be as valuable as the information recorded in it. This second point is valuable in establishing the environs of the first section—culture. There are many huge corporations in the world who have grown to eye-watering proportions, not because they have an ethically balanced corporate culture, but precisely because they *do not*. An organization that pays scant regard and wages to workers it considers little more than drones is not an organization that will benefit from any kind of analysis by this method. And any organization that wishes to grow in such a way has clearly read this far in a book which has little benefit for them.

Let us consider instead any other type of ethical, morally aware organization placed anywhere on Greiner's scale that wishes to understand, preserve and enhance its safety in all aspects of what it does. Whilst doing so, we should be mindful that there are no right or wrong answers to any of these sections. As we have caveated, the quality of the output from this undertaking will only ever be as valuable as the information input to it. As we have discussed with metrics generally, if one has an expectation of what the output *should* be, it is highly likely and perhaps unsurprising that that is what the output *will* be. Understanding the environs of all the functional control measures in an organization is key to developing better controls or, at the very least, being able to identify where potential failures in the controls might occur. The information we require for this endeavour is split into the following four sections:

Culture—Operations—Documentation—Administration

coda ▸noun *the concluding passage of a piece of music* ■ *the concluding part of a dance.*

(Oxford English Dictionary)

This forms, if you will, an assessment of the control behind the controls that, from an organizational perspective, shape the foundations of the flaws that we can expect. And although Andrew Sharman, in his book *From Accidents to Zero* (Sharman, 2016), stated that audits "have had their day", and even though the authors baulk at the term *audit*, we should remember the old adage: "what gets measured gets done". Of course, some risks—often those with the greatest severity or consequence—are not easily measurable, certainly in quantitative terms. The considerations detailed, therefore, in CODA are not checked for their *existence* but for their *suitability*. We should avoid the simplistic recognition that, say, *this* policy has been created, *that* study has been completed, or that *any* particular document is in place. To examine the environs of our control measures is to ensure that the policy actually represents what happens on the shop floor; that the study's conclusions were considered by the decision-makers and acted upon accordingly; and, that the document that is in place relays the right information to the right people in an appropriate way.

This reflects the *study* component of Plan-Do-Study-Act in that we must understand and measure our organization in order to be able to study it and proffer any recommendations for improvement. Whether one calls this an audit, assessment, analysis, investigation, or whatever is irrelevant. What is important is that it is done, and done properly. We call ours the case for safety.

> There comes a point where we need to stop just pulling people out of the river. We need to go upstream and find out why they're falling in.
>
> (Desmond Tutu)

Figure 6.4 from the publicly accessible standard PAS99, *Specification of common management system requirements as a framework for integration*, shows the interplay of the various functions within an organization (PAS 99, 2012). Although not unlike Porter's model, this diagram demonstrates the importance of the organization's function within the broader context of its undertaking, the inputs to it and the output from it, as well as other external forces. This correlates well with the earlier Figure 3.6. Our assessment of risk—and therefore the provision of safety

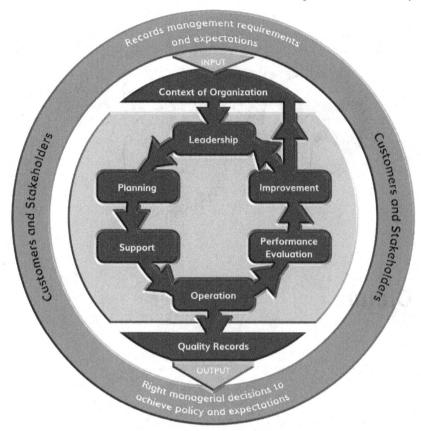

Figure 6.4 Structure of management system standards.

throughout the various functions—must consider this wider context, as we have discussed. External forces; inputs into the organization in terms of materials, personnel, finance, and so forth; outputs in the shape of products, services, deliverables, and the expectations of the customer—all of these will have some part to play in the proper identification of risk to the organization. In the following chapters we discuss these further in terms of what we should be looking for, and how they could impact on overall safe operation.

7 Culture

The culture of an organization can be influenced by many factors: its area of business; its size; its state of maturity (according to Greiner's model); its ambitions; and its senior management, owner, or founder, to name a few. Over the last two decades or so, we have seen the proliferation of ambitious, dynamic new companies resolving to change the way that business is done and often including the word "disruptive" in their mission statements or corporate testimonials. Disruptive has, until recently, only had a negative connotation; it has been seen as something to avoid, something contrary to the normal and acceptable way of doing things. But we see new ventures challenging the norms of business and gladly proclaiming that they are "disruptors", inspiring new ways of working, selling, and employing and paying people. This is to be applauded where it works well and can present new and meaningful benefits to all those involved in the organization. In cases where it does not work so well, it may be seen as a cover for ostensibly the same old organizational complacency dressed in exotic words and ideals. Often it is when ambitious young organizations become—due to their success—large, unwieldy organizations where the original dynamic hunger for change mutates into a relentless pursuit of ever-greater profit. The increase in the number of organizations that are self-proclaimed champions of change should be countered with

DOI: 10.1201/9781003296928-7

the ever-increasing distance between the incomes of those in charge of businesses and those who work for them.

There is a current fashion for referring to "safety culture", as if this should be in some way divisible from other aspects of an organization's function. Consider how likely it might be that a company could have an outstanding safety culture but a poor one in respect of employee engagement. Or an ambitious and well-documented safety culture but a lackadaisical approach to its customers in terms of the quality of safety of the products or services it provides. Quality, welfare, financial probity, management obligation, and engagement—all these, as well as many other factors concerned with an organization's undertaking, are a result of, or are influenced by, a commitment to safety and safe practice. Safety is an aversion to a harmful outcome, howsoever it may be derived, and culture is the key to encapsulating and practising that aversion.

Culture is predominately influenced by human behaviour, and it cannot be ignored that it starts at the very highest level in any organization, sometimes even in the societal norms in which the organization operates. The Korean Air cargo aircraft that crashed just after take-off from Stansted in 1999 is just such an example (AAIB, 2003). A relatively minor instrument fault, which could have been easily rectified by comparing the readout with two other similar instruments in the cockpit, was not picked up by the aircraft's pilot or co-pilot. In the few minutes that the aircraft took to take off and then bank wildly 90° to the left, there was no communication from the co-pilot and the warnings from the engineer and on-board alarms were ignored by the pilot. It is believed that the pilot, an ex-air force colonel, would have been so revered by the other flight crew that they would not have felt able to question his decisions. This deference is a trait of Korean societal norms and after the crash began to be challenged in Korean Air's training regime.

It is therefore important that culture must be monitored, controlled, and challenged throughout any organization's growth, especially by those who are responsible for setting it and correcting it. It must be monitored to ensure that what the organization says happens is actually what happens *at all levels*. It must be controlled to ensure that as the organization grows, adapts, and changes, that new inductees to the organization develop their thoughts and efforts on the accepted cultural doctrine *at all levels*. And, it must be challenged where the organization's culture is not being adhered to for whatever reason or by any individual. In simple terms, we must ensure that an organization can "walk the walk" as much as it "talks the talk".

7.1 Communications

Humans are phenomenally good at communication. We have developed highly advanced languages and methods of communication that have, quite literally, allowed us to explore the depths of the universe. But where there is great advancement there is usually complexity; and where there is complexity there is misunderstanding. Modern languages can have tens- if not hundreds-of-thousands of words, many with subtle variations of meaning. Languages are a reflection of history, environment, and national culture (consider the oft-quoted, but incorrect, "50 words for snow" attributed to the Inuit people, for example) (Kaplan & Larry, 2003). And our ability to communicate in any format has been radically altered by technology in recent years due to physical limitations on wordcount, the development of internet slang, and the cross-pollination of working by people with differing mother tongues and varying levels of understanding in languages foreign to them.

The irreconcilable differences in some technologies' abilities to work cohesively can also lead to difficulties. One program or application may not work on all types of hardware, or it may work with varying degrees of proficiency. Only a few decades ago, there were telephones for making urgent communications with people and handwritten or typed letters for more formal ways. We may have much faster and, some would argue, more flexible means of communication now, but are they to be considered as efficient if the message received by the recipient is misunderstood? Added to which is the question of the type of communication in terms of quantity (number of posters, signage, communiqués, etc.), quality (the style and content of the communication), and relevance.

7.2 Dissemination

We need to consider how information is moved between the communicator and the receiver, how often and using what systems, as well as whether these systems are appropriate to all the various types of communications. The systems of communication need to have a resilience built in to deal with the speed required for, say, a safety notice to be posted compared to a request for Bank Holiday cover in three months' time. How do different departments or even department heads communicate with each other and with senior management? And what systems are used to communicate with external influences on the organization: clients, customers, suppliers, creditors, and so forth? In terms of the supply chain, are they factored into discussions about growth, criticality of supply, alternative supplies, and their own supply and production issues; and if so, how, when, and how often?

The organization should consider its own stances on such matters as product integrity, quality, ethicality, sustainability, and modern slavery, and how this is communicated to its suppliers in the first instance and supplemental to any changes. As important is how feedback on the dissemination of all types of communication is, who collates this, and to whom it is passed. Sending staff an email newsletter every week is laudable unless nobody reads it, and demanding that suppliers only use x recyclable material in its product becomes merely a gesture if the deliveries from them are not checked appropriately. Similarly, an investigation following an accident for which there was a safety notice issued only days before should be enquiring as to the efficacy of the way that notice was delivered and to whom.

7.3 Reporting and metrics

It is easy for large organizations to get lost in a sea of data: information on productivity, quality, complaints, and sales—there are myriad nodes generating volumes of reports and statistics. Decisions about which metrics will be used and how they will be collated are often made early in an organization's journey, but assuring that they remain compliant with the standards set, and that they remain pertinent to a growing business, is crucial. Each department will probably provide data tailored to their own individual function, but when it comes to safety, there should only be one method of recording and collating data. Additionally, "safety" is often the first item on meeting agendas, outwardly giving the appearance of the value given by the organization to this important matter. In truth, it can have a deleterious effect by being "gotten out of the way" quickly before the "proper" matters such as productivity and profitability are discussed. Again, by considering all matters holistically with reference to safety (i.e., the aversion to harmful outcomes), its relevance becomes overwhelmingly one of natural, cultural interest, rather than a silo of data that receives cursory examination.

Understanding what needs to be the focus of an organization's attention, and what continues to be so, is allied to the need to record, verify, and review it. We have *planned* these things to be measured; we have *done* the recording; we should now *study* the data and its relevance and *act* upon our findings. This can be determined by examining previous decisions that have, to all intents and purposes, been made reliant on the data. Did senior management really listen to the data or was their decision a forgone conclusion? If it was, and something went wrong, is there sufficient humility within the culture to learn from this and improve?

7.4 Whistle-blowing

An employee raising a concern with their employer, where that concern is in the public interest, is protected in UK law by the Employment Rights Act 1996 (as amended by the Public Interest Disclosure Act 1998 (PIDA, 1998) from any form of discrimination in their work. And any confidentiality orders, gagging clauses, or non-disclosure agreements would be invalid in such cases. Yet still there are instances where individuals have received appalling treatment for having "blown the whistle", and organizations need to seriously examine the detriment to themselves in these cases. Saving a few pounds on a proper investigation or preventing the decisions of a senior manager becoming known is irrelevant compared to the cost of a substantial loss of business due to customers becoming aware of treachery in an organization and consequently turning their backs on it.

Issues within any organization need not necessarily be "in the public interest", but they will undoubtedly have some impact on its function. Even if proven to be incorrect, a proper investigation can demonstrate an honest, forthright culture. The method for individuals of all levels to make their concerns known, the protections on offer to them, and the way those concerns are treated and responded to are all crucial. There are too many examples where incidents have occurred with disastrous consequences for people's safety that could have been prevented if only an earlier whistle-blower had been taken seriously.

7.5 Feedback (lessons learned)

"We must learn lessons from this". How many times do we hear this on the news when yet another failure, fatality, or violation has occurred, often when a similar failing has happened in the recent past for which, on cue, the same mantra of "learning lessons" was brought out for an airing? In a perfect world, an organization would never need to learn lessons because the right decisions, involving the right people would have been made at the right time and disseminated to the right individuals in the right way. The possibility, however, of flaws occurring in any one of those events leads us to a real world where things sometimes go wrong. To err is, after all, human.

But our ability to learn lessons is hampered by two human responses: we do not like to be blamed, and we like to find blame in others. The latter is known as fundamental attribution bias and is the tendency of individuals to see others as more blameworthy than themselves. The "blame culture" has been strengthened by decades of a health and safety regime that has convinced itself and others that someone must be responsible for something having gone wrong. Human error, senior management, the board of directors, the client—anyone, it seems, at the top of the tree is culpable. And in many cases, there will always be some truth in this, to a lesser or greater degree. But accident investigation slowly began to realize that things go wrong sometimes of their own accord. It is not the event itself occurring but the multitude of fragile controls we have put in place which, occasionally, line up to allow harm to pass right through them. Whether you call this luck, resonance, or a one-in-a-million chance, it is clear that many times things go well and sometimes they do not. What must be rooted out of a culture is the knee-jerk reaction to find blame in any incident. The vast majority of individuals in an organization are, for the vast majority of the time, doing the very best they can within the circumstances to which they are exposed. If Fred has made an error causing loss, we must understand the circumstances *why* he made that error and correct them, rather than simply chastise or discipline him.

7.6 Behavioural maturity

We have seen that it is at junctures during the growth of an organization (which Greiner called "crisis points") that it can be at its most vulnerable as it matures to the next phase of its operation. This is perhaps less clear-cut in some of the "disruptive" business models that have spawned of late in an era dominated by the internet. These times of crisis are where important decisions about the organization's future are made and where lessons learned from the journey thus far can be implemented. In short, organizational behaviour matures as the organization itself matures, in size, capability, and awareness of safety. We see, however, that with some recent organizational examples, that behaviour does not appear to have matured at the same rate. This may be due to the "disruptive" nature of these types of organizations and their growth paths which, in normal terms, have been phenomenal to say the least. Some of the largest companies in the world have only been in existence for a few decades, in comparison to some of their stablemates who have existed for many decades, perhaps even a century or more. Modern business growth can outstrip our expectations and, therefore, even our planning.

The transition points between growth phases allow for a readdressing of the cultural status quo, and where new inductees to the business fill the new productivity requirements, there should be time allowed for "settling in" to the organization's standards, ambitions, and practices. We should note that these "points" of transition are not fixed but are, rather, indefinite periods through which an organization will pass (see Figure 7.1).

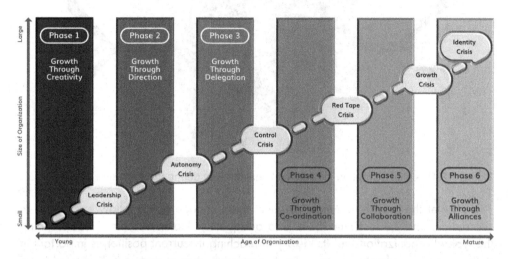

Figure 7.1 Adaptation of Greiner's model showing transition phases.

The period of time for approaching, transitioning through, and exiting each phase will depend on any number of factors, including the relative size of the organization. It can therefore be assumed that later transitional phases may take considerably longer to complete. Added to this period of time that each consecutive transitional phase takes is the disturbance to the normal function of the organization that transpires during each phase. The functions and outputs of the various aspects of the organization can—and will—be affected differently dependent on any number of factors. We discussed earlier the situation where the sales team brings in huge new orders, which the production and logistics teams struggle to fulfil. This inner turmoil can be made particularly difficult where communication is already poor, or where there is culturally a propensity to favour one area of the organization over another by senior management. The maturity of the interplay between the culture, operations, documentation, and administration going into a transition phase is likely to have defining consequences on their ability to "pull through" the transition and emerge, in the blinding sunlight of a new dawn for the organization, more robust and efficient. Figure 7.2 demonstrates how the previously ordered control of an organization can become disrupted and disturbed during a transitional phase and how this should return to normality. In this sense, it is beneficial to be prepared for the transition, the crisis point as referred to by Greiner, in order to establish a plan to deal with the inevitable disruption. This provides two useful outputs: firstly, we are planning for the worst to happen, and secondly, we have a plan to reflect on during a period of reflection once the transition phase is complete or near completion. This is akin to a "lesson learned" session where we can establish if our planning was suitable. We also have a template for planning for the next transition when it comes.

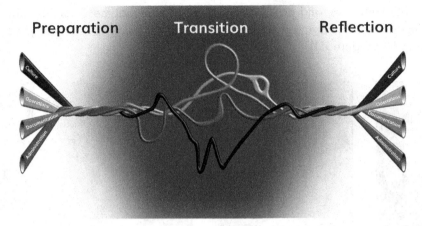

Figure 7.2 Potential disruption to organizational stability during the transition phase.

The faster the growth, the more frequent these change points follow each other in succession, and the less time there is to adequately deal with them. Understanding the type of organization and its history in reaching its current position is important in determining how effective is its ability to deal with change from a behavioural viewpoint. It is also important to understand that, as each transitional point is reached and progressed through, there is the added complication that there may be two competing systems running in parallel during this time. The disruption demonstrated in Figure 7.2 may result in the divergence of systems, and we have tried to emulate this in Figure 7.3.

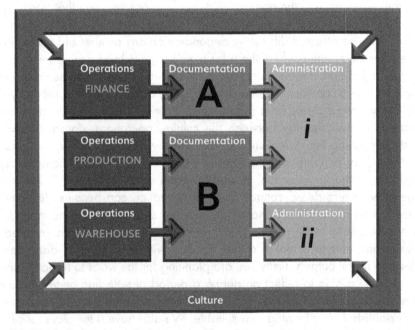

Figure 7.3 Division of systems during a transition phase.

Here we see how various functions of the organization might become aligned to differing documentary and administrative systems at the same time during transitional phases. These systems will be influenced by the original format as the organization entered the transition and those that it creates during it. The creation of additional or alternative systems may be due to the engagement of new personnel who bring with them their own experiences and ideas, or the organization attempting to promote new methodologies to take advantage of growth possibilities, or a mixture of both.

The change point is not, as we have said, a moment in time which is confronted, dealt with, and left behind in a blink. The period that each change (or transitional) phase exists for is x length of time, as is the period running up to, and out of, each change. These latter two periods are where the organization realizes there is an issue and begins to deal with it, and where it exits from this transitionary phase and settles down into its newly defined growth phase. An organization in the midst of one of these transitionary phases is likely to be increasingly at risk of lapses in safety as it is effectively "taking its eye off the ball" for x period of time as it internally struggles to redefine existing behaviours and methods of operation and adapt to new ones.

7.7 Supply chain

In addition to the culture that defines the relationship with a supplier, an organization's behavioural character also plays its part. Every business requires the supply of something or other to aid its undertaking, and the consistency of supply, quality, and price is crucial to its efficiency. And safety, also, is reflected in the consistency and quality of supply as they remove the possibility of having to source supplies elsewhere, perhaps at short notice, which also helps to retain the quality of the organization's own output. We should be considering the ongoing levels and content of discussions between supplier and the organization, and looking at whether suppliers are proactive in offering updates on any foreseeable supply issues. The method within the organization of circulating this information from suppliers is also of import—a supplier suggestion that there might be a reduction in supply of a material in two weeks' time is going to be of little use if the organization's internal mechanism for meetings means the procurement and production departments only meet every other month.

As organizational and behavioural maturity increase, there should be a commensurate increase in the cross-pollination of ideas and information between the organization and its suppliers. Designers within an organization should be discussing with suppliers about possible improvements in the material or equipment that they supply in order to bring about better products, and similarly, suppliers should be informing their customers of new or nascent materials or equipment that they are developing. Suppliers may also be aware of supply issues, shortages, or potential regulatory controls that may affect supply. Discussing these issues as early as possible helps to secure the supply chain and allows for changes in stock holding, handling, and manufacturing techniques to be addressed and implemented in a timely manner. This protects the safety of the organization's throughput process by ensuring that its personnel are aware of coming changes and can adapt to them. It also allows for proper testing of alternative products to ensure that they are fit for purpose and meet current quality standards.

It should be remembered that a workplace health and safety risk assessment should be reviewed in the light of technological change or development. Risk assessments are only as good as the current level of technology and operational understanding. One example is where evidence established in the 1980s that benzene (added to unleaded petrol along with other hydrocarbons) was carcinogenic, which led to new handling controls for workers exposed to petrol as part of their job (IARC, 2017). Another is the technical improvement in personal protective equipment for foundry workers, from being originally made of heavy leather, through aluminized-coated fabric to advanced lightweight and heat-resistant fabric. Fundamentally, we should be analyzing the supply chain to understand if procurement of materials and equipment is predominately one of buying the best for the purpose or of buying the cheapest for the sake of the shareholders.

7.8 Welfare

This has become another matter of great interest to the corporate world of late and is usually bundled up as health, safety, and welfare. Organizations have begun to realize that ensuring the well-being of their workforce can have many positive effects on morale, productivity, the reduction of lost time, and staff turnover. It is perhaps regrettable that it has taken the realization that welfare affects the financial bottom line for it to become of import to the board, but the fact that it has now is to be applauded. What needs careful consideration is, along with a commitment to safety in general, what lies behind the commitment to welfare? Laying out a few bowls of fruit in the office for a photoshoot for the latest website is hardly a commitment. Just doing enough to look after staff might be fine for a small growing organization where the individuals know that profitability is key to them having a job tomorrow, but in large organizations, the level and type of welfare might (should?) be very different.

Welfare of the workforce is inextricably linked to their overall safety, and this must have a genuine commitment at board level in order to set the agenda and, therefore, its culture. Welfare is also linked to dissemination (how are people informed of the benefits on offer), reporting and metrics (is the welfare of the correct type and level, and is it yielding the expected gains), and, of course, feedback.

7.9 Risk appetite

We discussed the varying levels of risk appetite that organizations might have in our previous book and how this can translate into its everyday function. In reference to safety and, moreover, safe operation, the appetite for risk is generally defined by the culture and the type of undertaking. What is important is that the control measures in place are suitable and sufficient to reduce harm to an acceptable level.

The acceptability of that level of harm is further dependent on whether we are working towards financial, regulatory, public, or reputational applicability. We have seen that many factors can affect both organizational and personal appetites for risk, as well as the slightly greater tolerance for risk, which we will discuss later. Firefighters, as we have discussed, have a large appetite for risk, or they would not be able to fulfil their role. This appetite, however, is countered by having many layers of control measures in place, such as intense training, quality of equipment, procedures, supervisory oversight, comradeship, and so forth.

We must also understand the *cohesiveness* of an organization's risk appetite; that is to say, the way that differing departments treat it within the general culture. This can have a bearing on safety implications, perhaps best illustrated by the police service in England. The police, like firefighters, perform an often inherently dangerous job for which they have numerous control measures in place. Successive governments have, however, reduced funding to the police service leaving them at precipitously low levels of operational personnel. Although this has no immediate effect on the types or levels of danger they may face day-to-day, it does critically affect one of their principal control measures; that is, having sufficient backup in the event of specific threats. This, along with other public services such as prison officers, illustrates the divide that often exists in private organizations as well, where the governing board has little understanding of the real work that occurs on the shop floor.

7.10 Training

Training, as we know, forms a fundamental component of competency along with skills, knowledge, aptitude, and experience (SKATE). There can be a tendency, though, to approach training as a "fit and forget" item. This may be perfectly adequate for simple induction training but may be damaging for more critical work practices. Currently, in the UK, first aid training is required to be retaken every three years—a moot point if the trainee has never had to perform CPR once in that time and finds themselves faced with a colleague who is not breathing, two years and eight months since their last training day.

Once again, it is important to understand that with training, it is a matter of the *intent* behind it as much as its quantity and quality. Does the organization treat training as part of its check-off of safety-related responsibilities, or does it engage with initiatives such as Investors in People? Are there sufficient first aiders to fulfil any regulatory obligations, or are as many staff as possible trained in first aid to promote safety and well-being in the wider community in which the organization sits? Do we understand if the training works as a "one off" or whether it requires a refresher or booster programme? Is the organization trying to train people simply to perform a function or is there a broader message about the development of empowerment amongst the workforce? As ever, feedback is important in understanding how effective training has been/is likely to be. This ties in with accident investigations in trying to establish if training was at fault (instead of assuming that the individual didn't "get" or understand the training provided). There also needs to be an examination of whether senior—and more probably, middle—management is as committed to training as it seems. People can often be wary of training those they oversee due to fears for their own job. This correlates intrinsically with the organization's culture, behavioural maturity, and attitude towards empowerment.

7.11 Organizational maturity

If behavioural maturity is about *what* decisions are made, organizational maturity is about *how* they are made. In the early start-up days of an organization, decisions can often be ad hoc and reactive. The eyes of the business are on the prize—that is to say, staying in business—in those early days of forging a place in the market. It responds to situations, issues, and crises as and when they occur, and this is all part of the process of learning from lessons. As organizations grow, the decision-making should become more orderly and organized, following processes and protocols sympathetic to the inherent culture. The model for the Capability Maturity Model Integration (CMMI) illustrates the various stages of this organizational maturity, as shown in Figure 7.4 (Chrissis, Konrad, & Shrum, 2011).

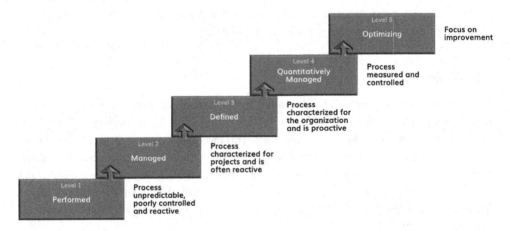

Figure 7.4 CMMI maturity model.

Although originally developed for software engineering purposes by the Carnegie-Mellon University in the USA and purposely refraining from mentioning "safety", the model is, like Greiner's, useful for graphically representing the various stages of an organization's functional maturity, allowing one to assimilate this with the various processes and procedures that are being undertaken. In terms of safe operation, if one were to replace the word "process" in Figure 7.4 with the word "safety", it can be readily seen that this model of maturity fits well with the perception and implementation of the safety function in various sizes of organizations. Of course, it could be speculated that, as an organization grows in stature, it becomes inherently more risk-averse, perhaps due to the increasing impact of the consequences of harm or possibly the effect of an ever-increasing number of documented processes that strangle creativity.

Regardless of this, the safety of personnel in any organization can be provided for *so long as the right control measures are in place*. In a small organization with a chaotic approach to process, the control measures might be the high levels of training, knowledge, and expertise that the (few) individuals in it have, combined with their empowerment to make decisions based on their skill sets. Much larger organizations might control their operations through tightly worded procedural documents and highly focused training modules for every task. It would also be necessary that the method of feedback should comparably increase in complexity and formality, and this requires ever greater efforts to ensure that this feedback mechanism is functioning as desired and that important information is reaching the right people at the right time.

7.12 Philosophies

The culture of an organization gives rise to certain ways—or philosophies—of dealing with functional aspects of operating a business. These philosophies are a significant indicator of how some of the benign, daily functions of business are approached and addressed. They can also help to provide some of the "colour" to a culture by either reinforcing or countermanding the keynote messages that the organization makes on safety. A business that declares its intentions on safety by relinquishing all maintenance to an external contractor, but then never follows up their work by monitoring or checking the results is an example. Culturally, the organization is promoting its commitment to safety by not exposing its own personnel to the hazard of maintenance. But its philosophy is flawed in not retaining at least someone who can verify that what they are getting in return is of suitable quality. And should there be a safety-related issue, both the contractor and the organization itself might find themselves suffering legal repercussions in such a scenario.

Outsourcing has become very much de rigueur in the manufacturing sector of late. Consider the number of Western businesses that have moved their production facilities to Asia, particularly China, over recent decades. This has seen the exceptional growth of products available for sale in Western markets at prices hitherto unheard of. It has also seen the profitability of some very large organizations grow exponentially. It could be argued, however, that this has simply moved the requirements of safety out of sight. One of the (many) reasons for the rise in the costs of production in the West, and particularly in Europe, is the cost of providing suitable safety and welfare arrangements for the workforce. But this is driven by regulation which in turn is in response to public demand. We do not wish to work in unsafe and dirty workplaces but, equally, we do not want to pay high prices for our goods. It is a paradoxical cycle that can only be answered by moving production away to where we cannot witness the safety procedures. This is not to say that all Asian production employs unsafe methods, but the disasters at Bhopal and the Tazreen Fashion factory in Dhaka, Bangladesh, perhaps give us an insight, albeit tempered with the knowledge of the failings of our own highly regulated industries. The latter of these two examples did also lead to the introduction of much better verification of overseas production facilities at least within the fashion industry. A belated acceptance perhaps of that industry's responsibility in a global marketplace.

7.13 Operating

The operating philosophy of an organization is principally driven by the development and adherence to its long-term strategy. This in turn tends to be influenced by its maturity and culture. A young, growing business might be battling on any number of operational fronts to survive its next year. A slightly more developed one might be looking to engage professional safety assistance, or even be thinking of taking on someone internally for the role. Large, established organizations will possibly have a phalanx of staff dedicated to developing long-term strategies across myriad areas of the operation. None of these are better or worse situations depending on our understanding of the way in which they are controlled and the relative measures put in place to effect that control. The small organization may, for instance, be using out-of-date equipment—a by-product of their financial situation perhaps—but if they have robust training and operating controls in place, suitable maintenance programmes in place, and a highly motivated workforce, then we might consider the risks suitably managed. A very large organization might have a ten-year plan for replacing their production machinery, but is this plan driven by productivity or price alone? Are the reports of the safety department considered in the procurement decision? Are the thoughts of the operators themselves taken into consideration? Does the organization even acknowledge that the operators may actually have an opinion?

In previous times, workplaces were often made up of many senior staff overseeing junior staff, who were apprenticed to the business, or learning through work. Staff turnover tended to be lower and people often worked for long periods for one company; sometimes for their entire life. We are not suggesting that this does not still happen to some extent today, or that it is a better or worse arrangement than greater transient freedom in employment. The modern working world tends towards much greater employment mobility and, with generally longer working lives, it is not unusual for people to have two or three different careers in their lifetime. We must not ignore experience in the SKATE equation however. Running aging machines in a factory is not a safety issue if the operators have "grown up" with them and know their foibles intimately, and the machines are regularly and properly maintained. But if this is the operational philosophy of the organization, and there comes a time when the operators begin to retire, then an issue arises as to who will run them safely. The risk of losing or gaining personnel is something that any organization is cognizant of; but in addition, it should also be considering how to replace the *knowledge and experience* they take with them when they leave.

7.14 Maintenance

An organization's philosophy on maintenance broadly falls into two categories: reactive and proactive. Is maintenance programmed down to the last detail, or do things get maintained when they stop working, or make unusual noises, or—for some organizations, worst of all—cause a fall in production? But this simplistic view ignores some important detail. Are the big production machines maintained regularly, as per the manufacturer's recommendations, but the staff toilets are left to disintegrate?

Are the delivery vehicles maintained by mileage or time? Is there a planned programme of maintenance for safety-critical components but not for those that feed them or are fed from them? An organization is a cohesive whole structure, and any programme of maintenance needs to be equally cohesive.

Another aspect to consider is that some devices require little or no maintenance, or would at least not cause a catastrophic safety issue if they were to fail. A computer, for example, is predominately a solid-state device that, should it completely fail, would possibly only cause a loss of data. There may be some spare computers in the IT department ready to be used as replacements, and that may be an effective control measure. If the computer is linked to the air supply regulators for deep sea divers, however, the controls will need to be much more robust. Other aspects of the maintenance programme to consider are how often it takes place and who makes that decision. Is it based on recommendation from the manufacturer, the maintenance contractor, an independent expert, or bitter personal experience? How is the maintenance verified and by whom? And what checks are made on the maintainer before they are engaged?

Inspections are inseparably linked to maintenance because these are the opportunities to, firstly, validate that the maintenance is being performed as required and, secondly, to identify problems before they become serious safety concerns. As with the maintenance itself, it is important to understand how inspections are carried out, their frequency, how they are recorded, and by whom, as well as the mechanism by which any possible issues are brought to the relevant person's attention. Carrying out detailed daily inspections which note the same fault repeatedly, without it seemingly ever being fixed, is a recipe for loss of morale on the part of the inspector. This invariably leads to those inspections becoming overtly routine and slipshod.

A planned programme of maintenance, for all relevant equipment and systems in an organization's arsenal, can be a highly effective safety control, and this involves not just the organization's culture and attitude towards safety but also the procurement department's. At the specification stage of any equipment purchase, there should be a complete understanding of the differential between its initial capital cost (CAPEX) and its ongoing operational cost (OPEX), which includes the maintenance of it. Purchasing the least expensive option does not always translate into greater running costs; nor does buying the most expensive guarantee the lowest. CAPEX and OPEX are, in some organizations, run along very different lines by different departments with different budgets, and this can cause a failure in the philosophy if it transpires that operating costs begin to rise substantially. Decisions to curtail or limit some maintenance procedures based on their expense will not have a positive effect on safety overall.

7.15 Management

The standard military operational thinking is that of setting a given number of personnel an objective, with an ever-greater number of collective objectives being divulged to progressively higher ranks of officers. This thinking was inverted somewhat by the creation of the Special Air Service in 1941 by Sir David Stirling, who realized the potential for very small numbers of soldiers to be empowered through training and trust to conduct highly tactical operations. Trust, empowerment, and training are powerful factors in a unified safety culture, but invariably only when the management of such a culture is extremely close to the theatre of operation. Having long, complicated chains of management can tend towards the disintegration of trust and the denial of the positive effects of empowerment by senior managers. The management style of any organization is undoubtedly connected to its culture, but individuals will also bring with them their own human traits and biases that can affect what the organization may understand to be a healthy culture.

If culture is the mind of an organization, its management style is its hands. It is how the culture is effectuated. The former must control the latter and, what is more, be assured that it *is* being controlled. The board must ensure for themselves that their predetermined management style is being accurately portrayed down the chain of command. As ever, the feedback mechanism is central here in ensuring that management issues are properly and effectively dealt with.

The management philosophy should also deal with how management progression is to be dealt with in relative terms. Is there a progression path from, say, operative to team leader to supervisor to manager? If managers are to be employed as the organization evolves, are they to be experienced in a specific proficiency, and if so, what would the baseline competencies look like? Management is linked to, and responsible for, more of the components of CODA than any other. The frequency and context in how management teams meet and discuss can provide vital clues as to how "in tune" the management style is, not just with the organization's culture but also those whom it manages. Do managers perform safety tours and inspections because they believe in safety as an all-encompassing benefit to the organization in general, or simply because their boss tells them to do so?

7.16 Procurement

We have already touched on how the conspicuous involvement of suppliers can have a beneficial effect on supplies coming in to an organization. The procurement philosophy will be based around this and should encourage engagement rather than just demand a sort of reverse sales psychology of haggling down to the lowest price. Beating down a supplier on price can be an interesting way to discover the novel methods that a supplier can deploy in order to save costs in the material or products that they supply. But procurement for an organization means dealing with capital expenditure as well as supply costs, and having a knowledge—as well as an empathy—towards the market, the equipment type and style, and (crucially) the ambitions and expectations of the organization, is vital. When specifying new plant, equipment, or structures, the organization should be able to compose a well-informed statement of requirements from which the procurement team can work. Expectations of budget, dependency, ongoing costs, and lifespan should all be defined. Equally, the procurement team should have the knowledge, experience, and empowerment to counter any over-ambitious expectations with factual arguments.

One key factor for the feedback mechanism for procurement is that of favouritism. It is an inevitable human trait and does not necessarily always have negative connotations. A particular supplier may be the favourite choice because of the quality of their work or material or products they supply. It may be due to the level of service they provide or because they have become intrinsically knowledgeable about the organization's operation. These can all be positive factors but must be consciously and conspicuously tested to ensure that they are not purely the result of complacency, familiarity, or inducement. Procurement is a cornerstone of an organization as it has control and management responsibility for material input into it, thereby having, vicariously, a hand in the responsibility of the safety throughput.

7.17 Risk tolerance

An organization's tolerance to risk stems from its risk appetite. This in turn will be influenced by many factors, such as its cultural and behavioural maturity, as well as the maturity of the risk management process itself and the sector in which the organization operates. It can also be affected by what the marketplace, its nation-state, or the public deem as "acceptable" levels of risk. An organization clearly must take risks in order to effect its undertaking. In terms of risk management, the term risk appetite refers to an organization's ideal for the amount of risk it deems "acceptable". Risk tolerance is the level of risk *beyond that appetite* that the organization is prepared to extend to. Figure 7.5, based on the Institute of Risk Management's risk appetite and

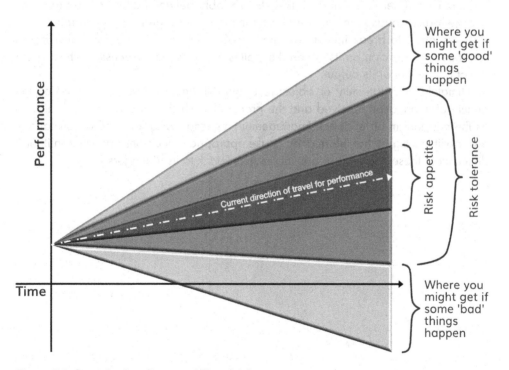

Figure 7.5 Organizational acceptability of risk.

performance diagrams (Crowe Horwath, 2011), shows how an organization's current trajectory of performance, or growth, over time is bounded by its appetite for risk. This appetite will affect, as can be seen, the increases or decreases in performance that the organization is *content* to accept. The organization's tolerance to risk is shown either side of this appetite and reflects the increases or decreases in performance that it is *able* to operate within. Beyond this are the areas where extremes of performance, either positive or negative to the organization's performance, may occur due to some tremendous event. These events may have such a profound effect either due to them being outside of the organization's control, or due to them not having been adequately considered or prepared for.

Often perceived as perhaps one of the more complex areas of overall operational management, risk is, nonetheless, a crucial factor in organizational expediency as well as a positive indicator of attitude towards safety. Simple statements such as "we will not expose the business to more than x percent of its capital in market investments" or "we will not deal with y types of suppliers" are an ideal starting point. As an organization grows and matures, however, these tolerances for risk may begin to alter. Very large organizations may take suppliers with whom they had not worked previously under their wing in order to provide education and guidance on becoming better businesses themselves. This may be as part of its corporate social responsibility (CSR) or as a reaction to a forecast of dwindling supply in the marketplace which must be addressed. As such, both the appetite and tolerance to risk should be dynamic—based on real-world information flowing efficiently into and around the organization. Tolerance will also likely be variable between different departments: a large luxury goods manufacturer, for example, may be prepared to sustain heavy losses in order to break into a new marketplace but will probably be unwilling to sacrifice any variation on its material supplies or production processes which could affect the quality of its output.

Ultimately, any statement or objective about tolerance to risk will depend on the quality of information received and the process by which it is disseminated, as well as the mechanism of feedback (management meetings, etc.), and the organization's own ability to digest that information, make appropriate decisions based upon it, and disseminate those decisions in a timely manner back to its managers.

8 Operations

If culture provides the "planning", then operations provides the "doing" of the organization. It may not be constructive to think of any of the components of CODA as having a greater significance than any other, certainly in terms of analysing an organization's real attitudes and motives towards safety. And it is true that an organization might not have a formally derived culture, and it may not have an advanced mechanism of documentation or administering its management function. But it will always be doing something operationally, even if it is just providing advice or lending money. Operations is where the undertaking is transformed from an idea into a reality and is, therefore, where the most immediate risks to safety occur. For this reason, it is where safety professionals tend to focus their attention. But as we have seen, the environs of the control measures that we introduce, in whatever capacity, have an almost infinitely variable effect on those measures. And it is for this reason that we should look deeper into the assessment of workplace risk and the consequent introduction of safety control measures.

Controlling work through risk assessments begins with understanding the process itself. We have discussed the use of job safety analyses to break down tasks into constituent parts, which can be individually assessed for risk, both within the part itself

DOI: 10.1201/9781003296928-8

and in the task as a whole. By addressing tasks in this way, it should be possible to extrapolate other operational factors involved, such as: who might be completing the task and their probable level of competency; where the task might be performed and whether this raises additional risks; how the task might be performed, perhaps during different times or seasons or weather conditions, or even under varying stressors.

8.1 Operational coordination

Coordinating operations goes beyond allocating shifts and forecasting production quotas. The culture and operating philosophy of the organization will of course shape it, but it should also have a built-in resilience to the natural vagaries that are constant in operations. Nowhere is it more significant to have an understanding of the operational environment than here, hence the reason that operational teams should always be invited to the design and specification meetings of new plants, equipment, and structures. Any new design is likely to have limitations of use prescribed to it, and it is these limitations that will have to be overcome by the operators should it transpire later that the design is not quite what it should have been. There have been warships designed and built for operating in cold seas against a particular (old) enemy which find themselves floundering in the warm waters of the Mediterranean due to a shift in global threats. And it is in these situations that the operators (that is, the crew) will be given the unenviable task of finding a solution.

Resolving issues at the operational level is of course considerably easier with the right information, the right team of individuals, and an empathetic level of support from senior management. The importance of *planning*, *studying*, and *acting* is no different in operations than any of the other components of CODA we describe; but they will almost certainly have a more immediate impact. The more ethereal components, such as culture and management, may change over time with the right attitudes and impetus; but if operations fail just once, there could be serious consequences for safety. Materials might not get stored properly; maintenance might not get done; production might not take place; people might get hurt. Therefore, coordinating operations involves providing the training, management, skillset, support, and dissemination of information necessary to provide the best foundation for any control measures required at shop floor level.

8.2 Operating environment

We have touched on the operating environment and how crucial this is to the operation as a whole. Whether it is a warship or a finance house—or an electrician installing a socket—*where* the task is carried out is as important as *how* it is carried out. (The example of the electrician is one we use often: consider installing electrical sockets on the ground floor of a house compared to fitting them in the nacelle of a 150 m-tall wind turbine—same task, very different environment.) The question, therefore, is: has the organization considered all the possible locations of operation, or is there a mechanism for managing this? If limitations of use have been put in place on equipment, how are these controlled? Further back from this, we should be looking at whether the operations department was involved in the design and specification of any equipment, and whether any previous changes in operational environment were factored into later designs.

The operational environment is also concerned with micro environments *brought to* the organization. A change to a supplier from the US, for example, might mean the pallet trucks in the warehouse won't fit American-style pallets. Or receiving a container from a country with an epidemic could render the receiving personnel liable to infection. Equally, maintenance is connected not just with use but with the location of use—such as lorries that work on construction sites or coastal areas which need increased frequencies of inspection and maintenance, and these considerations need to be addressed at the operational level if a full schedule of work is to be sustained.

8.3 Operating procedures

There remains some confusion regarding the terminology around operating procedures, and this can manifest itself fundamentally in two ways. Either there are not enough, or not specific enough, written procedures; or there is an overabundance of similarly styled documents, which creates confusion. Neither is a particularly palatable position to be in. Not having specific formal procedures against which to operate means that there is no way for the organization to gauge the performance of its operations against its philosophy or ambitions. There is no cohesive base for training new personnel on the operation, instead perhaps relying on existing experienced personnel to provide the training, which may be coloured with their own individual understanding of certain parameters of the job. And it will be less straightforward to identify what went wrong in the event of an accident and thereby learn lessons from the investigation.

Too much documentation can create obfuscation, which in turn can create a complacency amongst personnel by them ignoring the cascade of processes, procedures, work instructions, etc. Not being able to follow operating procedures because of their complexity or abundance will result in a situation akin to that of not having formal procedures at all. We find that as organizations grow in size and maturity, they tend towards an overabundance of procedural documentation. This may be due to the adoption of formal quality management systems, which sometimes confuse terminology or are themselves misunderstood as to what and how many operating documents are required. It may be due to the reduction of empowerment that inevitably comes with an ever-growing workforce or where departments become swollen with personnel without a commensurate growth in supervisors or team leadership.

Figure 8.1 demonstrates the relationship between various operational documents. A risk register will have been compiled, identifying all salient risks to the organization,

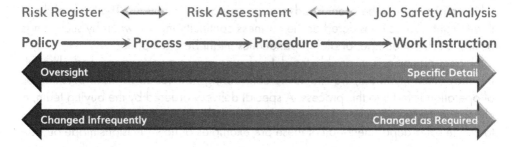

Figure 8.1 Relationship of operational documents.

both internally and externally. This risk register will inform both the risk assessments and job safety analyses that have been carried out (see Section 5.7). The contraflow of information at this level is important to ensure that risks and consequences are constantly identified, reviewed, and updated. This overall analysis of risk will inform the policy (or philosophy) of the organization and will set the tone of its expectations. The process states what needs to be done and why, in accordance with the ambitions of the policy. A procedure explains how that process will be undertaken, and the work instruction provides the specific detail to the operator on how to carry out the task. This is not to say that all tasks, nor all organizations, require this exact stratification of information. We must always bear in mind the phrase "suitable and sufficient". Operating procedures must be suitable for the task they are describing and sufficient for the number of personnel to whom they apply, their anticipated level of competency, and the environment in which they will be operating. Additionally, it is important that operating procedures are, like everything else in CODA, subject to an appropriate feedback mechanism in order to take account of changes in material type, production methods, staffing levels, environment of operation, technological advances, and any other factors that could affect the minutiae of how a task is done.

This feedback can be the most crucial, especially as an organization grows and develops. What was once accepted when staff were few and highly connected to the organization's founder may become less acceptable as the distance between senior management and the shop floor widens. Procedures may have been written by someone with a particular interest in productivity or overt safety. Later, an experienced operations manager might be employed who sees benefits in amending the procedures in order to create a more cohesively blended output. Or a highly skilled operator may be employed who has experience of the wider industry and proposes changes that will benefit the organization. Ignoring these sorts of proposals can create a situation again akin to not having formal procedures at all or where the organization finds itself in the position of believing it has appropriate systems in place, but they are, in fact, almost wholly bypassed at the operational level.

8.4 Planning

Operational planning extends beyond ensuring the right number of personnel are rostered to work each day. Planning for operations should initially include the three states of operation—normal, abnormal, and exceptional. That is to say, how is it to be planned and sustained during normal operation; when there is maintenance or repair taking place; and in the event of adverse or emergency situations. The latter of these three might also be considered as the business continuity model, whereby such things as fire, electrical outage, or global pandemics might be planned for.

Normal planning is invariably the duty of supervisory management, and although this may seem mundane, we should be considering how feedback from other areas of operation is fed into this process. A special delivery ordered by the buying team; a change in the specification of a supplied material; the demands of a customer agreed to by the sales department—all of these are examples of how decisions made outside of operations can have a critical impact on how operations plan for and maintain normal function. Abnormal operations introduce the same issues: has the maintenance department planned a deep clean of the offices on Tuesday; are all of the production machines set to be serviced on the same day; is the IT team going to disable the network next week whilst they install a new server? All these issues can be resolved through the simple process of having a communication mechanism that firstly ensures the right people meet and discuss the right information at an appropriate time, secondly, ensures that anyone affected by the outcome of any particular meeting who is not present is made aware of it, and thirdly, that special or exceptional issues can be raised promptly and effectively.

Exceptional planning will be most likely influenced by the organization's risk register. By identifying the possible risks to the organization, it should be possible to strategically plan an appropriate response to each one. This might include the members required to form a response team; the relevant departments that might be affected and how they will be communicated to; any individual who will take responsibility for the planning and dissemination of the response; the need for new safety protocols and how they will be introduced; and any physical changes that need to be made, and how they will be managed. These physical changes might be alternative power sources, places to work in, or ways of getting into the building. Planning should also consider any lessons learned from previous episodes of exceptional operation—what worked well and what did not work so well are key ingredients in developing appropriate responses in the future. There will certainly be more organizations in the world today who have a pandemic response formally written up in their continuity plan.

8.5 Emergency preparedness

Planning specifically for emergencies is somewhat different to planning for exceptional operational conditions. The very definition of an emergency means that although we may understand the causality of the emergency—be it a fire, flood, or structural collapse, for example—we may have little knowledge of its outcome or its impact. It should also, depending on the undertaking of the organization, extend beyond the usual Wednesday morning fire alarm test or a fire drill when the weather is warmer because people have complained about standing in the rain.

Health and safety practitioners often get maligned for asking endless "what if" questions: "what if this happens?", "what if that falls off?", "what if the lights go out?", and so forth. The frequent retort from those being asked such questions is usually along the lines of "well it's never happened before", as if substantiating that because it hasn't happened yet, it won't happen at all. This stems from the "we've always done it this way" statement, which is a kind of confirmation bias. If someone has never fallen off a chair whilst changing a lightbulb, why would they think it will ever happen? Fires, for example, in premises are, thankfully, quite rare but they do still happen and often cause serious injury or damage, and even fatalities. In 2017, the Grenfell Tower fire claimed 71 lives in one night, but in the year to June 2021, there were a total of 249 fatalities from fire recorded in the UK (Fire and Rescue Incident Statistics,

2021). This is in comparison to an estimated 1,472 fatalities reported on UK roads in 2020 (Reported Road Casualties Great Britain, Provisional Results: 2020, 2021) and 111 workplace fatalities in Great Britain in 2019/20 (Summary Statistics, 2020). Workplaces remain, thankfully, relatively free of serious harm, but that has taken decades of hard work, enthusiasm for change, and a raft of legislation to bring to bear.

Clearly, the object of assessing risk and introducing proportionate control measures is to prevent harmful outcomes from occurring in the first place. Fire alarms, property maintenance, and flood defences are all examples with which we are familiar as measures to control the more obvious risks. But global pandemics or a vapour cloud released from a nearby chemical plant may not be so readily controlled nor easily predictable in consequence. As with operational planning, emergency planning should be influenced by the organization's risk register, which in turn should be reviewed and updated regularly in order to stay well informed and effective. The register should identify those risks that are of peculiar importance to the organization. This may well include such things as supply line failures, marketplace closures (due perhaps to civil unrest, etc.), or extreme weather events. By accurately identifying—and quantifying— the risk and the consequences to the organization, the appropriate level of response can be formulated in an appropriate timeframe.

The policies, processes, and procedures for emergency preparedness are likely to follow similar lines to other operational documents, but the work instruction is likely to be replaced with an "action plan" for all individuals—operatives, supervisors, and managers—to follow in order to provide the correct response. Part of any action plan at a management level should also include the methodology of accident or incident investigation. Having a dedicated individual or team with the requisite skills, knowledge, and experience is valuable in providing a timely, empathetic, and measured response to any incident. It is important that the individual or team has the support of senior management and can present its findings in a non-partisan and unemotional manner in order that any recommendations are meaningful and pragmatic.

8.6 Emergency practice

It can come as little surprise that those organizations which operate in dangerous situations or environments tend to be the ones that make practising responses to emergencies a more regular and thorough affair. The military; firefighters; police officers; lifeboat crew—all of these face situations where the dynamic response to exceptional circumstances is crucial to their own and others' safety. The office staff in a bank are unlikely to meet with similar nerve-shredding situations, but we must still examine the environs of the control measures that are put in place even here. Are fire drills, for example, held regularly every third Tuesday in the month so that personnel clock-watch before the alarm goes off, making sure they are near to an exit? Or are drills held randomly and without warning to test the fire marshals' ability to shepherd their assigned teams to safety?

The institutions we have mentioned practise emergency drills because they are likely to face them often. In a normal place of work, emergencies may be—thankfully—very rare, but this can give rise to the issue of individuals not being aware of their responsibilities or required responses. We mentioned earlier in the book about someone with first aid training being confronted with a patient who needs resuscitation but who has not had to perform this in all the time since they were trained. Paramedics, on the other hand retrain on this technique often. Another example is that before flying, passengers are legally required to observe a safety briefing by the crew. But in the event of a real emergency, how many of those passengers would be able to remember precisely what they are required to do? In large organizations, it may be necessary to select a small number of personnel and train them more thoroughly than others to provide the necessary competent response. Simply sending half a dozen people on a fire marshal course for a day and leaving them to it is unlikely to yield the required amount of level-headed competence in the event of a genuine fire occurring.

Emergencies, by definition, are not ordered events and often have supplemental consequences: a fire cutting the telephone lines or a flood tripping the electricity supply, for instance. Practising emergency responses should consider the various vectors and velocities that any particular emergency could present. Again, this would be influenced by the findings of the risk register as to which events *could* occur, their possible consequences, and any supplemental effects that they could have. Having a fire drill when the seat of the fire is designated at one of the fire exits or having a windowless training room full of inductees on their first day when a flood trips the electrics are simple examples. This in turn will help to inform the current control measures and establish if further controls—or variations of them—are required. This feedback should travel all the way back to the risk register and policies and then return to inform the procedures and action plans for everyone to be made aware of.

In some large, hazardous organizations, where the Control of Major Accident Hazards Regulations (COMAH, 2015) apply, there is the responsibility of not just the organization but also the local authority, under whose jurisdiction they fall to extend their considerations to neighbouring locations. In reality, this may be a statutory requirement for only a small number of organizations, but it is still a worthy consideration for all types of undertaking. How would the emergency services attend; what access or egress are they likely to need; how and where would the organization's personnel be kept safe during the emergency, as well as any neighbouring personnel? All these additional factors should be considered when practising the various responses in order to ensure that effective controls are in place.

8.7 Contingencies

Contingency planning may well be an outcome of identifying threats to the organization through general preparedness planning. One of the most impactful contingencies that perhaps organizations should now be considering is that of another pandemic. Businesses around the world had to implement protocols and practices in a very short space of time; practices that until only a few years before would have been unheard of.

But contingency planning will almost certainly include more than just physical changes—with growth comes the challenge for organizations to adroitly transmit information to relevant individuals, teams, and individuals about those changes. This may require setting up specific groups to handle such aspects as personnel, deliveries, production, safety, and so forth. If decisions need to be made (and undoubtedly they will in dynamic situations), there should be a protocol for how and by whom those decisions are made. Contingency planning should also consider how the contingencies will be communicated if, indeed, the communication systems themselves are compromised: if the computer network is down, for example, or key decision makers are away on business or holiday.

8.8 Continuity planning

Beyond contingency planning is asking the question, "how does the organization continue in the event of a major disruption?". It may, of course, not be possible to fully function, but with the administrative function of most organizations being conducted electronically, there should be some ability to carry on to some degree. Continuity planning may seem more associated with the financial safety of a business than personal safety, but we should still be considering the personal aspects of any such planning. If personnel are to be laid off for a time, what, if any, financial arrangements can be made for them; if production can be moved to another location, how will personnel be relocated there; and what facilities—toilets, canteens, washrooms, etc.—might be able to be provided, are some examples. Equally, for some events, there may be the need to provide some form of well-being support for staff who were involved in or witnessed some disastrous event. Many organizations contract occupational health advisors or may even have them as employees. We should consider if these advisors have the relevant skills and knowledge to counsel employees who may be suffering from emotional trauma brought on by what they have been involved in.

Continuity planning may seem the ultimate example of planning for the worst and hoping for the best and, as ever, should take its cues from the risk register as to what extremes of operation an organization *might possibly* be exposed to. It may, however, also be an ideal opportunity to ensure that existing control measures are suitable and sufficient, and if deemed not to be so, then plans can be put in place to enhance them. These plans should similarly be informed by the likelihood and severity of any risks, as well as the dynamics of any possible consequences. (The velocity of risk is the speed at which it can increase in severity: a flood, for example, could almost immediately curtail an organization's ability to receive deliveries, store materials, and make and deliver products; a strike in a foreign port might, on the other hand, take several weeks to begin to affect production. A risk's vector is another way of looking at how a risk can "travel" or affect other aspects. A fire, for instance, could release noxious fumes, catch adjacent buildings alight, drift smoke across a nearby road, cause reputational issues, disrupt power supplies, and so on.)

8.9 Maintenance and repair

Tied in with the maintenance philosophy culture is the plan for how systems, machinery, buildings, and so forth are to be "brought back on line" following some major event. It may seem highly speculative, especially when one cannot be assured of what type or level of maintenance or repair is possible, if at all. There are, however, events and consequences of those events for which some level of predetermined response can be planned. An outage of the computer network, for example, will invariably need a resetting of the system, along with checks performed on the integrity of the data. The aftermath of fire or structural collapse might be, up to a certain level, appropriate for the organization's own building maintenance team to clean up. The organization's regular electrical contractor may be able to restore power and confirm the integrity of the electrical system.

There are numerous companies which provide business recovery services, but is the organization likely to call the first one they find the details for or one that they have predetermined has the right levels of resources and competency? Are the regular contractors for repairs and/or maintenance able to cope with all forms of recovery, or should there be alternatives in place for larger or more drastic situations? If specialist contractors are kept on hand to deal with potential outcomes, are they checked regularly to ensure that they still offer an appropriate service or, indeed, are still in business? Is their competency pre-checked, such as qualifications, relevant institutional memberships, and insurance cover?

9 Documentation

As an organization becomes larger and more complex, it is inevitable that it will require ever greater numbers of documents. And the "paperless office" we were promised by television science programmes of the 1980s, with the advent of new-fangled devices called "computers", does not appear to have abated the need (nor desire) for voluminous amounts of paperwork. Irrespective of this and of the way documents are controlled—be it from a simple spreadsheet or complex, bespoke software—there are essentially two important criteria for a documentation system. Firstly, it should be adequate and appropriate for the task and the number and variety of documents it is designed to manage, and secondly, it ensures the right information is relayed to the right people at the right time.

Document systems, like many aspects of organizational growth, can be hampered by an organic rather than a concerted development. And the adoption of standardized management systems such as ISO9001 (ISO 9001, 2015) or PAS99 (PAS 99, 2012) is unlikely to quell this growth; on the contrary, they tend to exacerbate it. This tends to be through a combination of Parkinson's Law (where the pursuit of implementing a management system creates a department all of its own) and where a system's requirements are taken too literally. Certainly, the authors have worked with

DOI: 10.1201/9781003296928-9

client organizations whose obsession with "complying" with every detail of some management system has led to an abundance of documents that inevitably become unwieldy, poorly understood, and rarely used. Except, of course, during an audit.

The proliferation of documentation since the promises of moving away from it may also be as a result of an increasing appetite for litigation over the same period. Certainly, the introduction of the Health and Safety at Work etc. Act 1974 in the UK, with its unique condition that someone prosecuted under them must prove their innocence rather than have their guilt proven may lead employers to ensure plenty of paperwork is completed in order to substantiate their position. The reason for having to prove one's innocence is straightforward, of course. If a person falls from height from an unprotected edge clearly something has gone wrong: the evidence is in the falling. The individual or organization controlling the person who fell must then demonstrate that everything that could have been done was done to prevent it. But if someone has already fallen, it would seem extremely unlikely that it was.

Paperwork is not everyone's forte, and obviously, not every career path requires a predisposition to either enjoy it or understand its true purpose. This knowledge—that some people are good with filling out forms while others are not—is an example of an environ of one of the types of control measures that are used to provide a safe workplace. Introducing endless forms for workers to fill in, knowing that they are unlikely to be completed either on time, properly, or even at all, is an admission that that measure is already unlikely to yield the levels of control we expect of it. And threats of discipline or sending emails full of wrath are also unlikely to achieve positive results, certainly in the long term. People who are threatened to complete something they do not understand nor care for will find all sorts of ways to complete the task without actually completing it properly. Forms will be filled out with spurious data, signatures, dates, and figures—all to keep the boss happy. And when that information is fed back into the management system, apparently showing how everything is "on target", senior management will have no inkling of the trouble that could be fermenting on the shop floor.

9.1 Documentation management systems

As organizations grow and develop into sprawling administrative colossi, they often develop complex systems and subsystems of documentation to appease the myriad departments and functions it has. Fundamentally, it is irrelevant what these systems are or how they operate: what is important is that they are cohesive and complimentary rather than oppressive and divisionary. The latter two characteristics can develop out of a poor culture where department heads vie for attention and affection—or simply engage in traditional chest-beating about whose department is more important and, therefore, most worthy of the glossiest documentation system. As we have said, standardized management systems may create the perception that every function or procedure needs an array of primary and supporting documentation, but this is rarely the reality.

Another issue with complex document systems is where there needs to be cross-referencing. One long document might refer to several other long documents to create the sense that "everything has its place, and there is a place for everything". But if one of the documents that is referred to changes its document or version number, the original will now be referring to something that is out of date or non-existent. This may continue for years before a sudden, unplanned event—like an accident—highlights the fact that the organization's supposedly carefully curated document system imploded a long time ago. Creating a complex document system is not the issue—creating endless, complicated forms really is not hard; creating short, meaningful ones appears to be much more difficult. Maintaining that system to a high degree of accuracy, on the other hand, certainly is and can become more apparent where documents are used both electronically and physically. Has one department printed a thousand copies of a document to use for some purpose, only for that document to be revised halfway through using them? If that document is subsequently transposed onto an electronic system, is it now in the wrong order, or does it contain outdated information?

When looking at the document management system of any organization, we should consider if its policy of use is suitably robust and mature. This should state, at the very least, how the system operates, by whom, what the decision-making process is, and who is the ultimate authority. It should also consider whether everyone who needs access to it has the appropriate level of access. Are editing controls set to those who should have them, as opposed to those who perhaps should not? Where there is cross-referencing, are updates accurately performed and in a timely manner? Are printed versions of documents suitably annotated to warn of confliction issues, or are there print run controls in place? When it comes to drawings—for design or research purposes, for example—are they controlled in terms of revision numbers, and are the new revisions disseminated to everyone in the design or research team accurately and in a timely way?

Feedback from the users of documents is also important in their development, and this process of feeding back user experience is always worthy of consideration. Documents are often developed by people who may not actually use them, and therefore, the process of developing any document should be undertaken like any other design project. That is to say, accurately identifying what is required of the project's output; putting forward possible solutions; discussing those solutions with the relevant stakeholders; and developing the final product in line with the defined statement of requirements. That way, the users get the document they need, and the data it generates is relayed appropriately to the recipient. Feedback on documents that have not undergone this process is extremely valuable in setting the scene for designing replacement documents in the fulness of time.

9.2 Documentation management

With a suitable document management system in place, it is then the work of the administrative function to properly operate it in relation to the culture that the organization has determined. Or, put another way, it must be effectively used in order to provide and disseminate the information the documents contain. This ties in with the systems' policy we described earlier and, in this sense, how the policy will be policed.

> *Procrastination is probably my worst habit out of many. Especially when it comes to paperwork.*
>
> (Tom Conti)

We have already discussed that keeping on top of paperwork is not everyone's specialist subject, nor are many people employed specifically to do so. The management of documentation should therefore accept this and accommodate it accordingly. Are there, for instance, arrangements in place to help mitigate this, such as accountancy software that allows people to photograph their expense receipts for submission rather than list them on a form? Novel approaches like this are welcome, so long, of course, that they receive an appropriate level of attention in terms of standardization and usefulness.

Document management should consider essentially the same matters that we discussed under *Operations*. That is, the environment in which documents will be used, in the sense of both the physical and ethereal environment—where is the document to be used and who is it designed to be used by, and who will it inform? We should consider the procedures for completing documents in terms of the time of completion, frequency, level of detail, and so on. And we should consider how documents will be moved, transmitted, stored, accessed, altered, and archived.

9.3 Document version control

We have touched on the issue of version control, the vigorous administration of ensuring that only the latest documents are accessible and in circulation. This is one of the most crucial aspects of any document management system and yet is perhaps the hardest to satisfactorily guarantee. It is particularly important during a design process, where drawings and specifications might be updated regularly during the early stages as the design matures. We advocate the importance of safety-related decisions being made at the earliest opportunity in design projects, and it is, therefore, crucial that these decisions are based on the most up-to-date information.

Today, there are numerous digital systems to record and disseminate drawings to a project team. Again, it must be stressed that the *construct* of the system itself is irrelevant, it is the way the system is operated and administered that is crucial. Bear in mind that many, many safe products have been delivered throughout history without relying on any form of complex, integrated (and possibly expensive) software. Our earlier reference to the Supermarine Spitfire is a case in point: a highly effective fighter aircraft delivered at a fast pace in exceptional circumstances by a large number of designers working with paper drawings and set squares. Complex does not always equate as clever, but efficiency is certainly always the key.

Similarly, safety-related documentation should be equally regarded as a crucial component of any document management system when it comes to making sure only the latest versions are in use. And let us be clear in our use of the word "safety". This may refer to a risk assessment that has been updated to show a new working practice as much as it may refer to a prequalification questionnaire for a new supplier that now includes a section on modern slavery, for example. Safety is a function of an entire organization, whether it relates to health and safety, financial probity, operational safety, supply chain assurance, and so on and so forth. The effectiveness of control over document versions is a key component of standardized management systems, something that many organizations already comply with. By ensuring this vital aspect of document management, we see that this has ramifications for safety as well.

9.4 Policies

Something that is often witnessed in organizations of all sizes, not just large or complex ones, is the voluminous size that general policy documents can run to. With health and safety policies, this can frequently be the case of a "more words must equal more safety" approach which, unfortunately, is rarely the case. Obviously, there is a legal requirement for a specific health and safety policy (from the Health and Safety at Work etc. Act 1974), but this does not restrict an organization from incorporating it into other policies where safety has some bearing on other aspects of its undertaking. Just as a risk assessment does not have to be restricted to purely health and safety issues—it can encompass all safety-related issues if properly laid out.

The drawback with overly worded, cumbersome policy documents is getting the people who should read it to actually do so. And, moreover, understand it. If a workplace has a high proportion of employees who do not have English as a first language, for instance, it might be reasonable to assume that a complicated and lengthy document, quoting this and that regulation, is unlikely to be well understood by a sizeable proportion of the workforce. That could present a serious flaw in any control measures that are necessary to comply with the policy's ambitions. One of the amendments to the accreditation process for the ISO14001 standard in its 2015 iteration (ISO 14001, 2015), was that sample individuals should be interviewed regarding their knowledge of the organization's policy on its environmental management system. This is useful feedback to identify how well the organization disseminates its ambitions to the shop floor rather than just developing esoteric policies and documents to appease the auditor. But how often does the safety department go out and ask people on the shop floor, construction site, or office what they know of their organization's health and safety policy?

In examining the policies for any organization, we should be considering whether they are, first and foremost, relevant to the organization and suitable for its undertaking. There are numerous generic policy documents available to those who choose to use them, but it should be borne in mind that an organization that does not follow its own policy is already in breach. An "off-the-shelf" policy is very unlikely to capture an organization's *actual* ambitions or requirements. Suitability will depend not only the undertaking but also—as we have seen—on its applicability to those who should be aware of its content.

We should also be considering whether the policy is reviewed periodically to ensure it is still fit for purpose and amended in a timely and appropriate manner when it is found not to be. Amendments will also (obviously) need to be disseminated accordingly. As we discussed in the section on version control, any policy that refers out to other documents, especially where document version numbers are quoted, should be managed robustly to ensure the references are accurate. The periodicity of review, the mechanism of amending and disseminating the policy, the system for updating references, and the process of checking the policy's implementation should almost certainly be included within the policy itself.

9.5 Procedures

Processes, procedures, work instructions: the terminology preferred by any particular organization for the document that states how any task is performed is less crucial than the need to actually have one in place, as well as how suitable, pragmatic, and accurate it is in context. In health and safety parlance the work instruction is often referred to as a method statement. And once again, there is often a disconnect between this and the work instruction or procedure, as though something that defines a task for the purposes of health and safety is somehow different, or ancillary, to "normal" documentation. Clearly, there is an absolute need for the output of a risk assessment to directly influence the document that illustrates how any task is completed. The two are not mutually incompatible.

As with any documentation, a procedure should be specifically aimed at those who undertake the task, and we should therefore be considering their competencies, levels of training and understanding, and the level of empowerment granted in the tasks being undertaken. The categories identified in Bloom's Taxonomy (Bloom, 1956) may provide a useful indicator in this respect. We should also consider how a procedure deals with tasks either side of it—is there a requirement for some level of integration, for example, or the checking of any input or output to the task? The mechanism for reviewing the documents, and the making of any requisite amendments, should be defined in the policy, which in turn should be identified in, and controlled by, the operational and administrative functions.

Exceptions should also be dealt with by any procedure, where relevant, illustrating to the operator what to do in the case of any event that is outside of the normal parameters expected of the undertaking. And if any exceptional output requires referring to other documents or procedures, these should be clearly identified and the references updated accordingly if they themselves are amended at any stage. Similarly, we should be considering any hazardous inputs to the task—in the form of materials, processes, or equipment—and whether these have been specifically identified in safety-related documentation, which, as we have stated, should be fully integrated with instructions for completing the task. Under UK health and safety legislation, there is a requirement to complete a risk assessment, for example, under all of the following:

- Management of Health and Safety at Work Regulations 1999
- Manual Handling Operations Regulations 1992
- Personal Protective Equipment at Work Regulations 1992
- Health and Safety (Display Screen Equipment Regulations) 1992
- Noise at Work Regulations 1989
- Control of Substances Hazardous to Health Regulations 2002
- Control of Asbestos Regulations 2012
- Control of Lead at Work Regulations 1998

It may not be necessary to complete all these as individual assessments for any given task, or they may be completed along more general lines with regard to the organization's undertaking overall. If separate risk assessments have been completed, however, and are relevant to any given procedure document, how they are referenced to it should be considered.

One aspect that can be invaluable to any procedure document is brevity. And, moreover, brevity in plain English. Long policy documents full of management phraseology is one thing, but overly complicated instructions for completing a task are unlikely to be (1) read and (2) understood. This can result in a self-fulfilling flaw in one of the most important control measures available, that is, the document that actually explains how to perform a task safely. Simple language is usually a much better way of defining the parameters for a task too; one that generally allows fewer errors through misunderstandings or violations. And, should errors or violations occur, it is generally easier to discipline against simplified language. We discuss parameters for tasks further in the delegation section of the chapter on Administration.

9.6 Certification

Certification may range from the statutory—such as a Thorough Examination required by the Lifting Operations and Lifting Equipment Regulations 1998 (LOLER, 1998)—to the voluntary—such as an accreditation under an ISO standard. Authorizations are another type of certification that are important in identifying an organization's commitment to safety and quality. Any type of certification is generally the result of a not inconsiderable amount of effort and time invested by any number of individuals in achieving it. It is entirely understandable, therefore, that organizations often post copies of accreditations to this or that standard on noticeboards around their facility. Less often seen, however, are the certifications connected to safety, such as the thorough examination we have mentioned, test certificates, authorizations, and so on.

In construction, for example, the authors' main area of work, we often see in refurbishment or demolition projects that electrical wiring is partially removed by qualified electricians, but that there is no disconnection certificate in the vicinity to advise other workers of the safety—or otherwise—of any number of dangling cables. In the Piper

Alpha disaster, we know that a permit to work certificate was in place but that it was not suitably posted to advise the next shift of the potential danger. Displaying the correct certification is as important as having it in the first place.

Certification can, unfortunately, fall into the same pitfall that we see in connection with other documentation, in that safety-related documents do not get the same publicity that, say, an accreditation in a quality standard does. On the face of it, this makes no sense. An organization that proudly declares its commitment to quality but eschews its relationship to safety—in all its guises—is surely, not *wholly* committed to that quality ethic. Safety and quality are inextricably linked, and the former will always improve the latter whenever properly executed.

10 Administration

In CODA, we take administration to mean, in the primary definition of the Oxford Dictionary, "the process or activity of running a business, organization, etc." One could say that it is the "acting" part of PDSA but that would ignore the fact that administration (or management) is the cornerstone to making sure that not only every other part of the organization's function gets done but that it gets done in the correct way. Having a vibrant culture, lean and efficient operations, or a bespoke, integrated software document package is unlikely to provide the expected returns without proper administrative control over it all. Administration is also often found to be a root cause of failure during accident investigations, where control from the strata of management proves to be inadequate.

It could be argued that, during the transition phases of an organization's growth that we have described, administration is the function that is most likely to be stressed and tested. This may be due to several factors, but we would postulate that it is primarily due to just one, the human factor. An organization's culture may change during a transition phase to one that is more compliant or empathetic; or the operational set-up may need amending to improve productivity; or documents may have to be altered and updated. It is probable that we have all seen documentation

DOI: 10.1201/9781003296928-10

that has hand-written additions until such time as the form gets revised or reprinted. None of these are likely to cause acute issues for the organization, even if they form part of a greater, chronic problem that creates holes in the environs of its control measures.

But administration failing because two individuals refuse to speak to each other, or a manager who cannot see the true value of safety as a function, or someone who denies responsibility because "it's not my job" are examples of immediate threats to safety controls.

Rasmussen may have defined violations in human behaviour as being routine, situational, or exceptional, but he may have been able to add obstreperous as well. But perhaps we should not be too harsh on the human species. It is certain that humans are prone to making errors, but they are also capable of making important decisions with limited information and according to their own skill set. Of more importance is that we have known this for a very long time, long before psychologists and safety system experts started to formulate models of behaviour. The proper function of the administrative system should be to accept the frailties of human behaviour and make the necessary adjustments for it. If we state that "you must wear gloves when performing this task" as a control measure, we must prepare for the inevitable question "why should I?" as a potential flaw in the environs of that particular control.

Case Study—Clapham Junction Rail Crash (1988)

The collision of three trains just outside of Clapham Junction station resulted in 35 fatalities and nearly 500 injuries. It was caused by errors made during a signalling upgrade that had been carried out several days prior to the crash. During the rewiring procedure, one wire had not been stripped back correctly and another had not been removed as required by the work. The wires caused a short, which led to the light signal failing to display a red "stop" signal after a train had stopped on the line. Although this was the immediate cause, it transpired in the subsequent inquiry that the technician who had made this error had been promoted to his current position despite never having undertaken formal qualifications. This was contrary to a stipulation made by his employer, British Rail.

The technician had not been trained on, nor was he aware of, the documented method of conducting the work. Nor was his supervisor on that day aware of the documented procedure that required such work to be checked. The area engineer was also unaware of this checking mechanism and did not believe it to be his responsibility.

The technician's poor standard of work had been a constant factor throughout his 16 years working with British Rail. In all that time, it had never been challenged by any of his supervisors, nor had he received training in the correct procedure. This continued through to the Testing & Commissioning Engineer responsible for the work who was himself untrained and poorly supervised by senior management.

Additionally, the rostering of men to complete engineering work on the railways was calculated using out of date manpower levels from 1986. Levels in 1988 were significantly lower, and therefore there was a great reliance on overtime working, resulting in technicians and supervisors having very few days off over protracted periods.

10.1 Organization

The organization of the administrative function is usually and effectively dealt with by way of an organigram (or organizational diagram). The most useful of these, certainly in large organizations, is where there is an overarching organigram which is "zoomed in" at each departmental level to portray ever more detail. One crucial aspect of organigrams is keeping them up to date. A beautifully designed diagram complete with everyone's photograph and job title is not much use to an inductee if half the people on it left over a year ago. Another useful supplement to the diagram is identifying who is responsible for any particular team, department, or individual in the event of their absence. With UK holiday entitlements, one only has to reach a level of eight employees before technically someone is absent on holiday every day of the year. Having prior knowledge not only of who is alternating for someone else but that there is already someone programmed to do so, can have a positive impact on the morale and enthusiasm of those whose immediate leader has suddenly gone sick. Even if that alternate person is just "holding the fort" until the cavalry arrives.

The cultural ambitions of the organization can be implied by an organigram—the reporting lines may appear dictatorial or they may promote inclusivity to the viewer—but also the diagram can promote lines of communication, in line with the communications philosophy. Simple diagrams, however, are unlikely to provide all the relevant detail on how the administrative function is arranged, especially for new staff.

10.2 Delegation

The delegations within the organization should be clearly and unambiguously defined. The "general manager" of a small organization is likely to be responsible for just about everything, but with organizational growth and development comes the likelihood of more and more departmental heads, supervisors, and team leaders. The responsibilities for these roles and functions need to be defined not only to provide the individual with formal parameters for their job but also to provide the organization with evidence if an individual strays from, or fails to complete, some component of their job. Delegations (or proper job descriptions, if you prefer) should also be cross-checked to ensure that proper delineation of responsibilities is in place, which will further enhance both the aforementioned benefits.

In many operational activities, such as driving forklift trucks or operating machinery, it is commonplace to have a formal authorization, signed by a manager, for each individual who is suitably trained. Again, this allows for proper control over who can operate what and is also an excellent cross-check against competencies for various equipment and tasks. Formal authority for carrying out administrative functions is much less common but would have equal benefits. It should be recognized at the senior management level, however, that although it is possible to delegate responsibility, it is not possible to delegate the duty of care that an organization has for overall safety.

Another advantage of ascribing individuals' responsibilities in a precise, formal manner is that their skills, knowledge, and training can be readily identified. This makes it easier to see if and when additional training is required: a manager who has never been responsible for personnel might need conflict resolution training before joining the sales team, or an operative being made up to supervisor might benefit from a management awareness course, for example. It is also possible that an individual who has been with a business for many years, and who has seen it grow considerably in size, is possibly going to have various responsibilities due to their time served and inherent knowledge of the organization. It may well be that should they leave and be replaced by someone else that that person would be unable to "fill their shoes" so readily. Understanding the precise expectations of the role provides clarity for everyone.

10.3 Work structure

Getting the right people to complete the right tasks at the right time to a specified output may seem a rather obvious ambition for any organization. It could be argued that that is indeed their raison d'etre. A company that makes widgets, and has always made widgets, may see little need in formally structuring their work. But what if a client asks for a different material to be used for their widget? The value of the order is large enough not to be ignored, and the new sales manager is full of enthusiasm to make their mark on the old-fashioned owner of the business. In a small organization, this might be resolved over a coffee in the sales office; in a larger organization, with its strata of management and complex lines of communication, it may be a little harder. Certainly, just allowing the production personnel to "figure it out" is not an option.

The work structure should also show both directions of travel: do issues with production follow the same path down which the work came, or is there a "short circuit" for issues to be directed straight to the quality team or senior manager? This is also connected to levels of empowerment within the organization, again part of its cultural setting, which may allow flexibility within the structure for abnormal and exceptional circumstances to be raised more quickly or more accurately to the most appropriate person. Complimenting this is the need to identify any alternate individual who is responsible in the event of absence, as well as any key decision makers whose opinion needs to be sought in the event of an exception to normal production. This may be someone different depending on, perhaps, a particular volume of production, exceeding a certain monetary value, or the importance of a particular client.

10.4 Training programme

Clearly, it is important that in any organization individuals have the correct type and level of training for the processes that they need to follow, the equipment they must use, and the strategies they must employ in order to complete their tasks competently. The type of training required for any job role will be influenced by the culture's training objectives and the operational coordination as much as the individual needs of the work tasks themselves. Beyond that, however, is ensuring that any form of training is completed in a timely manner, as well as the substantiation that any type or level of training remains effective.

We should be considering the system by how the need for any type of training is identified; the efficacy of the rollout to relevant individuals; who undertakes the training, where, and how; and how any training package is validated as meeting the set objectives. This last item can be assessed as part of any incident investigation, whereby the effectiveness of any training involved is juxtaposed against decisions (or omissions) that may have led to the incident occurring. To be clear, we are speaking of all forms of incident where a harmful outcome occurred: that is to say, an unintended loss of production, financial loss, or injury.

Retraining rates are also to be considered, especially where high-risk or complex tasks are being undertaken, where technical advances are likely in the process or the equipment used, or where the operational environment of the task is subject to change or variability. Most of all, we should understand if the ambitions of the organization, in terms of training needs and provision, filter down through the organization in a commensurate way.

10.5 Communications

The administration of communications will cover a wide variety of forms of communication—verbal, written, electronic, planning, signage, reporting, investigating, and so forth. But in essence, the only thing that really matters is that communication is effective and timely. The organigram we discussed previously may well be used to indicate general lines of communication but these should also be considered within the role descriptions too. Organigrams, however, rarely deal with exceptional forms of communication, like those required in an emergency, a grievance, or for whistle-blowing. These types of communication invariably require some form of "short circuit" to the standard model, and how these are effectively communicated (i.e., that individuals are made aware of them) and administered is extremely important.

Communication is the cornerstone of human interaction, and how and when it is done is crucial to every organization and every individual within it, as well as every other external organization with whom it deals. Having an ideal mechanism for communicating is only effective if it is properly managed and disseminated. It also needs to be subject to scrutiny in order to ensure it remains fit for purpose. As with so many other facets of CODA, we should be able to identify communication failures in any incident investigation. Equally, with safety tours or inspections of the organization, we should be able to ask without fear or favour whether the communication system is appropriate. Instead of asking operators why they are performing a task outside of accepted norms, we should be asking if they have had sufficient training and communication on the task. Was it written in the right way for them to understand? Or even in a language with which they are more comfortable? In safety terms, we often see extensive and prosaic risk assessment documents that may deal more than adequately with identifying risk and recommending controls, but are they actually relevant to the people they affect?

This links in with the metrics and reporting that we discussed previously. The administrative function of the organization needs to address the issue of whether the metrics being reported on are a true reflection of the organization's ambitions or simply a self-praising assessment of a list of objectives that were likely to be achieved anyway. Senior management, in this respect, needs to be fundamentally aware of what they have demanded from the reporting system and how this is communicated back to them. Any and every system needs to be tested, and communications is no exception.

10.6 Procurement

In procurement as a cultural function, we discussed the need for a wholesome working relationship with suppliers in order to bring about a multitude of benefits. To administer procurement properly, we need to accept two primary truths: that everything has a cost and that everyone needs to make a profit. Certainly, there are organizations that like to play one supplier against another in order to save a few pounds. And save them they may well do, but at what cost in the long term? A far more mature response is to enter into dialogue with suppliers where both accept these two truths and discuss openly. Procurement is not a football match where one side has to be seen to win. It should be a mountain-climbing expedition where everyone aims to reach the summit safely in one piece.

A procurement culture that is driven by squeezing suppliers on price at every opportunity is fine so long as it is accepted that some of those suppliers will either scrimp on the quality of what they supply or will inevitably go out of business. Either situation will undoubtedly have a consequential effect on the organization's various departments including, ultimately, safety. The authors have worked with many clients who dispassionately opt for the least expensive in any procurement situation, despite numerous historical issues that have been reported to them when having done so in the past. Issues such as equipment no longer maintained because the supplier went out of business; or quality issues with production due to inferior materials. Ensuring strong, mature dialogue with suppliers probably starts with empowering procurement personnel to work cohesively with them. The proper management, training, and communication with those individuals will be the necessary control measure to ensure that empowerment is well-founded.

10.7 Supply chain management

The management of the supply chain is the administration of regular inputs to the organization, such as raw materials, maintenance and repair functions, cleaning operations, and so forth. In some organizations, this may be undertaken by the buying or procurement department, or some aspects may be undertaken by them, like raw materials, for example. This may occur where an organization buys materials on the spot market, dependent on price and availability. Maintenance and repair contracts may be set up by the procurement team to then be managed by the operations department going forward. How the supply chain is managed will be entirely dependent on what best suits the organization.

Ensuring that it is managed, however, is vital in maintaining control. Very large organizations with multiple sites around the country may choose a maintenance contractor, for example, that offers country-wide response capabilities in order to minimize the number of local contractors that the organization has to deal with. This may make perfect practical sense to procurement and logical sense to the operations team. But if the level of service at site level is not as effective, pragmatic, or helpful as it could be, then the savings in cost are likely to be outweighed by hidden cost factors such as inconvenience, loss of production, reputational losses, and so on. Even at local and national government level, we frequently see contractors falling short of their promised service levels. If the operational staff on the shopfloor are complaining about service or quality levels, and their complaints are not being addressed, then they will eventually stop complaining—and this is often taken as an indication by the contractor (and the organization employing them) that "all is well" after a few "teething issues". Effective management of the supply chain needs to deal with any issues robustly; but more importantly, it needs to be *seen* to be dealing with them robustly.

The use of pre-qualification questionnaires (PQQs), such as the publicly accessible standard PAS 91 (PAS 91, 2017) for the construction industry, is useful in determining a potential supplier's credentials prior to their engagement. They can be used to provide a like-for-like comparison between competing suppliers upon which either a decision to award a contract can be made or perhaps for arranging a shortlist for further consultation. They should be regarded, however, like an MOT certificate for a car—they are valid only at the time of completion, and they certainly do not expose any issues that might be hidden below the surface.

As with the procurement function, the most successful method for ensuring effective control over the management of the supply chain is through communication. It can also be achieved in the first instance by presenting supplying contractors with a statement of requirements of what the organization *actually* needs, as opposed to what they believe they need or have always had in the past. Does machine maintenance have to take place overnight—at costly night rates—or is there a contractor who can minimize disruption to production to the point that it becomes cheaper to do it on day rates? Is there a moratorium on a particular type of raw material being supplied because of a historic issue with quality, or is there now a technical solution that, if implemented, outweighs the investment cost?

There also needs to be a dialogue about how any supplies to the organization, or contractors entering the organization's workplace to carry out work, are controlled from a safety perspective. Are there storage issues with materials, for example, in terms of humidity, temperature, or proximity to other types of materials? Do maintenance engineers need to have areas demarcated for their own or others' safety whilst they work? And how is this information communicated to the individuals and teams who will have to carry out these controls, and how are they assigned the necessary authority, empowerment, and equipment?

10.8 Asset management

Small organizations tend to have direct knowledge of the assets at their disposal—machines, vehicles, tools, handling equipment, office equipment, and so forth. They tend not, therefore, to have asset registers. As they grow to medium-sized organizations, the senior management knows it is something they need to do but all available staff are busy with their own functions, and the organization does not have the financial capacity yet to employ someone just to go around recording assets. By the time the organization is large enough to be able to employ a clerk solely to put together an asset register, it has so many assets spread across myriad departments, locations, and disciplines, and those assets have been disposed of, replaced, and upgraded so often that keeping track of it all can be a full-time fool's errand. Hence we find so often that the asset register is unlikely to be a creditable account of an organization's condition.

Of course, keeping an asset register is a valuable way to begin to satisfy the requirements of the Provision and Use of Work Equipment Regulations 1998 (PUWER, 1998), which apply to any equipment supplied for use at work. The guidance to these regulations is clear that they apply to almost all work equipment, although specifically not livestock, substances (for example, chemicals or concrete), structural items (stairs, walls, fences, etc.), and private cars (L22, 2014). Although applicable to virtually everything used in the workplace, the regulations are clear that a risk management approach needs to be taken in assessing risk and applying controls. Hence, a punch press will attract more interest than a screwdriver, for example. The guidance is also clear about such things as considering specific groups of individuals who may have to use any equipment, instructions for people who do not speak English as their first language, any relevant training or required competency, and so on. All of these matters we have highlighted throughout CODA.

It is sometimes the case that organizations have organically developed asset registers for individual departments. The IT department may have one for screens, laptops, and PCs in order to keep track of equipment they have issued to staff. The workshop may keep records of hand tools for replacement purposes and the production manager might keep one for machinery for their maintenance records. Electrical equipment is most likely to be recorded on a portable appliance test (PAT) record. In essence, it does not matter if there is not one whole cohesive asset register, so long as all assets are recorded somewhere and with a commonality of information. And most importantly, the location of each register is readily accessible and in a common format appropriate to the whole organization.

Recording and tracking assets properly will integrate with the ability to identify any competency needs, replacement timescales, or maintenance requirements. This, of course, can only be achieved if the recording of assets is integrated with their general management control and the flow of information between different departments. We need to consider how the HR department is to book training sessions for new machines being delivered, how procurement is made aware that all the canteen kettles are beyond their serviceable life, or how the maintenance department is told how to operate the replacement fire panel that has just been installed.

Therefore having an appropriate, complete, and up to date asset register is fundamental to not only establishing proper regard for the legal responsibility of providing safe workplace equipment but is also an incredibly powerful tool in creating a foundation upon which many other aspects of the proper and safe function of the organization rest.

10.9 Risk management

The management of risk is, as we have discussed at length, vital to the safe operation of an organization in toto. The output of the risk management process is likely to be a multitude of risk registers, assessments of threats and opportunities, policies, procedures, and control measures. The administrative function will be concerned with how all these documents and data are reconciled, disseminated, and, crucially, acted upon. Having an elaborate and well-informed risk register is of little use if the actions and controls it recommends are never implemented.

With the administration of risk management, we should be considering firstly how the register of risk is compiled and by whom, and additionally, under whose authority. It is not uncommon to have several registers that different departments or disciplines in the organization complete for their own purposes, but this can lead to some of the risks identified in them being isolated from those identified in others. Financial departments might, for example, create a financial risk register at the behest of the financial director in order to provide insight into market analyses, consumer patterns, investment opportunities, and so forth. If this is not reconciled with the register for safety risks, however, the financial department might be unaware of the issues that this identifies. Operating without the correct insurance, exposing operatives to dangerous work conditions, or investing in poorly specified equipment might, for example, bring about costs to the organization in terms of claims and fines that could radically alter the finance department's balance sheet.

In this respect, it would be very easy for us to state that a risk is a risk, irrespective of how it is perceived to affect an organization. But this would be to ignore the unfortunate truth that most organizations, and indeed, individuals, still see safety—and particularly, health and safety—as a function all on its own, dislocated from the "real world" of business. As we demonstrated in *An Effective Strategy for Safe Design in Engineering and Construction* (England & Painting, 2022) and, hopefully, as we have demonstrated in this book, safety is inherent. It is a component of every facet of an organization's undertaking and has a profound, if not the most profound, effect on it.

Additionally, in the process of risk identification and recording, we should be considering how the information gleaned from the risk register(s) is then translated into action at shop floor level. And by translation, we mean not only the mechanical method of informing people but the reasoning too. Put up a sign next to a push button that reads "do not press" and see how many people press it. Similarly, telling the buying department that they cannot buy a particular material might not prevent them from doing so if they come across the deal of the year and buy a warehouse full of it at a knock-down price. Telling them, instead, that the engineering department has identified a risk that feeding the wrong material into the machines will render them useless will reinforce the reasoning and prevent them from trying to do their best for the company by attempting to save a few pounds.

10.10 Actions management

Actions derived from a risk register should be accompanied by the name of the individual who will champion the action—that is to say, drive the action to make sure it is completed. Note: "make sure it is completed", not actually complete it. If this champion is a junior staff member, perhaps because no one else wanted the responsibility, then it might be difficult for them to be able to prompt more senior staff to get the action done. This is where a robust mechanism for administering the management of risks is crucial: if senior management does not want to be labelled "champion," then they must accept, with grace, that they will in all likelihood be receiving cajoling (if polite) emails from junior staff.

It should also be considered that risks are rarely static. As one is dealt with, another may appear somewhere else; or risks may change over time or inevitably as the organization grows. The actions and control measures required will, therefore, also change and evolve. Documenting these changes and disseminating the amendments they might bring is also an important function. Having a dedicated individual to track all the actions and changes might be appropriate, or at the very least, it should be a topic of discussion at senior leadership team meetings. Making sure that actions and control measures are kept high on the agenda is also an important part of the feedback loop in terms of ensuring that risks remain identified, controlled, and above all *understood*. Asking the "what if" questions is always useful in this respect.

As important as any other aspect of the administration of risk is the proper documenting of the outcome of meetings, decisions, and discussions surrounding the subject. As with any design risk assessment, recording the reason that certain decisions were made and why could prove invaluable in the event of a harmful occurrence. Establishing that, despite the organization's best intentions and efforts in controlling a certain risk, something still went wrong could be invaluable in the face of enforcement action or litigation. And if something going wrong results in an accident, it is equally important that any resultant investigation is reviewed in connection with the elements of CODA to establish if changes to any of the supporting arguments should be made. For example, if we support our case for safety through the evidence of training, the argument for which is the implementation of a training register, an accident investigation may highlight that this register is not as complete as it should be. Let us imagine that the HR manager has left and the post remains vacant for a period of time, allowing the register to become out of date. Analysing this information against the environs of CODA, we might determine that having only the HR manager in control of the register is an unnecessary additional risk. The administrative system is subsequently changed to ensure that two or more managers authorize entries to the register and progress has been made.

10.11 Safety management

We have stated our dislike of safety being treated as a ring-fenced function of any organization's undertaking, treated as it so often is as like a silo, into which a collection of thoughts, ambitions, and individuals are placed in order to stew in a stultifying gravy of regulation. It may seem, therefore, at odds to list safety management as a separate item in our list. But, whilst safety—and, moreover, health and safety—is treated as such, we must deal with this accordingly.

To begin with, safety *must* start at the very top of any organization. This is a cliché that everyone knows, but it must be clearly demonstrated, not simply talked about or signposted. During accident investigations, it often becomes clear that there is a divide between what an organization *says* it does and what it *actually* does in terms of safety management. The administrative function of safety should act as the conduit for not only disseminating the ambitions of the organization through the various strata of management to the shop floor, but also to receive information back from them and submit this to the senior management. In order to do this effectively, safety management must rely on all the factors we have considered so far in CODA, which are:

- Culture
- Operations
- Documentation
- Administration

Culture

Essential in establishing the ambitions of the organization with regard to safety. It should reflect the cultural safety aspirations of the entire organization and not isolate a "safety culture" from it. Safety—that is to say, the effective management of the risk of harmful outcomes—is inherent in all functions of an organization's undertaking, not solely what is perceived as the health and safety aspects of it. From these cultural ambitions it will be possible to derive policies and procedures to implement it. And here it is important to remember that what is stated in the policy is what is reflected operationally. If there is a disparity between the policy and the operation then one or the other *must* be changed in order to reflect this. The culture should also establish what monitoring and testing of the culture is to be applied and how the mechanism of feeding back results from these monitoring functions. We need to consider that if control measures are not functioning as required, how is this identified, recorded, and submitted to the appropriate individual or department. We should also consider how that individual or department responds, and how the cycle of disseminating information back to the sender is achieved.

Operations

The layers of operatives, team leaders, supervisors, managers, and so on should have a cohesive system through which to receive and submit information about the implementation of control measures. Whether this is through meetings, emails, an intranet, or whatever, is irrelevant—what works for the organization, and can be robustly sustained, is the important feature. Operations is often at the sharp end of an organization in terms of realizing new and emerging safety-related risks, hence the need for timely and accurate information about these risks being transmitted. The administration of safety management will also be concerned with the responses to exceptional and emergency situations, and it will invariably be the responsibility of operations to deploy the correct response. This is also true of the maintenance and repair functions that should prevent many such events from occurring in the first place by identifying and eliminating poor operational conditions. Ensuring strength in the operational structure will help to get relevant information to the right people at the right time, as well as ensuring the feedback on any inadequate performance is reported accurately.

Documentation

There are those who decry the amount of paperwork that they perceive is generated by the "safety culture". In some respects, there may be some truth in this, although this may be more to do with the predilection of organizations to treat safety as a standalone matter rather than an inherent factor in all its functions. Clearly, too much documentation can be as bad as too little, and finding the right amount is, for most, a matter of trial and error. Quality, environmental, operational, and human resource management systems all tend to create their own administrative document frameworks, and differing departments will tend to apply their own priorities in completing them. But the integration of safety into all of these will create better results in every function: safer operations will lead to better productivity; safer working practices will lead to fewer accidents and less lost time; safer control on supplies will lead to more standardized quality of output, and so forth. We should be considering the availability of documentation; who it is written for and how it is to be used; how it is controlled, updated, and verified. And, as always, we should consider how documentation is reviewed in terms of its effectiveness via a feedback loop and by investigations into accidents and incidents.

Administration

The administration of safety should not be considered something that is controlled by one person or department with the word "safety" in their title. Safe operation should be the *only* standard form of operation that any organization should strive for. Not one based on chasing elusive numbers or endlessly reporting vital and yet meaningless statistics. Safe is the opposite of unsafe: it is the adequate and considered control of risks: risks that will always be apparent in any organization and in any undertaking. Safety management is knowing what risks are evident, having the necessary control measures in place to deal with them—within the restraints and ambitions of the organization—and having the relevant support mechanisms in place in order to put and keep those measures in place, review them as necessary, and modify them as and when required. And behind those control measures lie their environs, the subset of controls that influence and determine how effective the measures are. Those environs are culture, operations, documentation, and administration: *CODA*.

11 Maintaining CODA

Having come this far, the casual observer may have noticed that there has been very little specificity in our proposals for what to look for and how best to manage threats to control measures in an organization. There has been no talk of what roles to employ for, what documents to create, or what systems to adhere to. The confident reader will understand that this is entirely deliberate. The use of the word "consider" has been of paramount importance throughout. There is no magic bullet; no defined "golden thread" to follow to bring every risk or worry to an end. Just as there is no specificity in Plan-Do-Check-Act, nor event tree analysis, nor in most health and safety regulations. Nor, for that matter in the old-fashioned Policy, Organization, Planning, Implementation, Monitoring, Auditing and Review (POPIMAR) that controlled safety thinking for so long.

11.1 Observation of hazards

The observation of hazards and risks pertinent to an organization and the consideration of how those hazards might affect it is nothing new. But the unified, cohesive identification of hazards—*all* hazards—and their influence on the safe operation of an organization is. We have seen that the influence of seemingly unrelated matters can have devastating outcomes on all aspects of safe operation, despite there often being what we might refer to as a "safe system of work" being in place. But these are the flaws in the control measures that we have examined—the *environs* to the controls we so deliberately put in place to effect safe operation. There is no simple solution to understanding this, and every organization will have a unique set of operational parameters, environmental concerns, and external influences to contend with. CODA provides a template for thinking about and identifying these, but it will take skill, knowledge, and experience to fully understand how these environs challenge safety in any given organization.

Imagine a small factory as an illustration of an organization (see Figure 11.1). The building rests on foundations and it has doors through which material (such as raw stock, data, chemicals, and so forth) enters and a finished product emerges. The building has walls, floors, windows, and equipment to bring in, maintain, repair, and operate. And it has people who will not only work there but also visit for any number of reasons. Imagine now that the foundations of the building are the culture on which

DOI: 10.1201/9781003296928-11

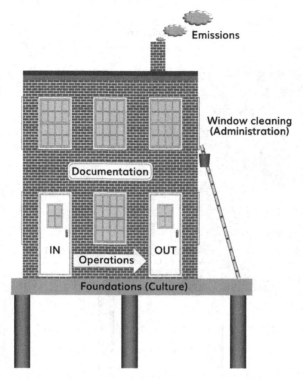

Figure 11.1 Small factory representing organization.

the organization is built. They must be robust enough to support the whole structure and be of a type that is appropriate for the prevailing environmental conditions. Buildings have been erected on rock, soil, marsh, sand, and everything in-between because they had suitable foundations for the ground type. Similarly, organizations have been built up in industries such as aerospace, engineering, finance, construction, and many, many others because they had systems in place commensurate with their particular industry.

The flow of material in, and product out, of the building is the organization's operational aspect. If the operation grows or changes so the "building" must be adapted to suit. Otherwise it could result in a mountain of raw goods in the car park, or the machinery might be making millions of the wrong type of product. The walls, floors, and windows of the building are the administration of an organization. They must properly reflect the organization's ambitions to visitors, and there must be enough light being let in so that people are not working in the dark. And flowing around the building is electricity and data that carries information to and from interested parties and machinery to keep everything moving along. That flow is the documentation that any organization needs to operate effectively.

Imagine the building growing by being built upwards (see Figure 11.2). We might suppose the foundations will have to become stronger and that the doors will need widening in order to accommodate additional throughput. The walls will need to be

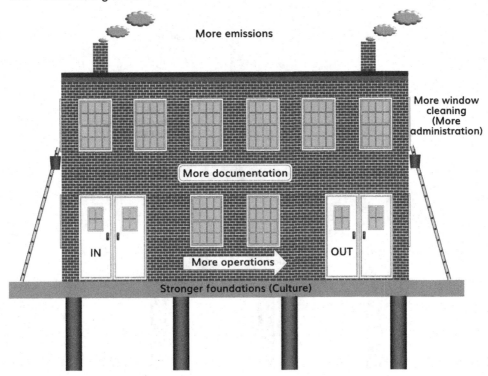

More emissions

More window
cleaning
(More
administration)

More documentation

IN

More operations

OUT

Stronger foundations (Culture)

Figure 11.2 Large factory representing organization.

reinforced, and the cabling system will have to be enhanced or possibly stripped out and replaced altogether. In the same way, growth in an organization will require a stronger culture to keep a larger number of employees advised of its ambitions or because it is now trading in many different markets. Operations will have to be altered to cope with greater production—with all that that entails—and administratively, the organization will have to adapt and strengthen to cope with the additional "load" of a greater throughput. Finally, the documentation will have to be enhanced to deal with longer chains of communication or new pathways of information that flow in, out, or around the organization.

So we have examined our "building", checked the foundations, the walls, and the wiring—what is next? Feedback. We have discussed at length the importance of feeding back information gathered about the organization and, in particular, the environs surrounding the control measures that may have perhaps created an unfounded sense of security, especially among senior management. Auditing against a predefined standard is a good way to ensure that the questions raised by the audit are answered, but this is of little use if the *wrong* questions are being asked. And by wrong, we mean questions that might simply ensure that we comply with the audit rather than seek out the real issues that could cause real harm. Feedback must be sought out without fear or favour, and it must be accepted by the organization's senior management as a genuine and objective appraisal of its actual implementation of safety. One method of collecting feedback is the safety tour, where a senior manager walks the shop floor, engaging with every level of employee in open discussion about their role and the organization's control of it. It is important, though, that a safety tour is undertaken without prior expectations or a tightly defined schedule. The senior manager needs to engage in conversation and be able to inquire earnestly about the matters that confront them, without constantly keeping an eye on the time. An open mind and a keenness to understand are the best, and perhaps only, tools that any senior manager needs to take with them on such a tour.

11.2 Maintenance of information

Another important aspect—perhaps *the* most important—is the maintenance of all the information we have gathered using CODA. We must not complete our study and understand its findings only to then close the file, place it upon a bookshelf and let it slowly gather dust. There is a reason that the instruction manual for a car does not alter throughout the duration of the car's lifetime but that the same car must undergo an annual safety check and service. A car's controls and driving aids do not change, but the forces that the car is exposed to whilst being driven can alter radically. Similarly, an organization's policies may not change from one year to the next, but the influences it is exposed to *can* alter. The marketplace in which it operates; the regulations it must abide by; the number and type of employees it engages; the materials with which it makes a product: all of these, and others, are subject to change at various times and with varying velocities. We must, therefore, be continually identifying the risks, and consequences of those risks, to which any organization is exposed. CODA is not a standard, nor a linear path: it is a cycle of continuous observation and understanding.

Writing a risk assessment or creating a work procedure is all well and good, but they are not going to be effective if the risks identified change over time, if the operator performing the task never reads the procedure, or finds a better way of performing their task without telling anyone. Too often—and certainly in the case of some of the disastrous events that we have covered in this book—the fact that some document or other, or some system or other, is in existence belied the fact that in the background the environs of that document or system were flawed, cracked, or failing. We have discussed the reason for our preference for the plan-do-*study*-act motif originally envisaged by Dr Deming rather than the more common plan-do-*check*-act. It is because we should be studying the results of our observations; understanding them and adapting as necessary. "Checking" implies that we merely ensure that something has been done and this has, perhaps, been a failing that has gone on for too long.

We have also seen over the years the laudable ambition of many a senior management or organizational strategy meeting to "put safety at the top of the agenda". This, also, can have an unfortunate and deleterious effect in placing safety (and more frequently referred to as "health and safety") as an agenda item that can be "got out of the way" quickly so that the meeting can get on with the real business of doing business. It was never meant to be so, nor—we are sure—is it meant to continue to be so. But the connotation is there and it certainly encourages those who view safety as some sort of "bolt-on" to operations. It is hoped that this book has made clear that safety is a factor in every function of any organization's undertaking: in operations, finance, administration, documentation, employment, training, sales, the supply chain, maintenance, production, and so on. Any of these functions can influence safety to some degree or other, and the fact that any of them are themselves influenced by any number of external and internal factors also gives rise to the importance of understanding those influences better. Here are a few examples:

Finances: Financial strain on an organization might result in less maintenance being carried out (rendering machinery less safe to operate); cheaper raw material being bought that doesn't have the same storage requirements; or cheaper and less effective personal protective equipment being procured.

Supply chain: A wide, uncontrolled supply chain might expose the organization to variable quality raw materials; materials may be sourced from corrupt or oppressive regimes; or it could be subject to extreme variations in supply (a "feast or famine" situation, which could also affect finances).

Employment: Growth in sales might require a sudden upturn in employment which outstrips the training capabilities of the organization; there may be a decision to hire exclusively foreign workers who do not have English as a first language (with subsequent issues for signage and documentation); or there may be a round of redundancies that increase the workload and pressure on those who remain.

Training: The training regime might become out of date in comparison to the operational requirements of the latest work practices, machinery, or quality standards; training may be outsourced to a contractor who does not fully integrate with the organization's culture or ambitions; or a long-term external contractor might go out of business at a crucial time of expansion in the organization.

We are sure the reader may be able to think of many more examples where the interconnectedness of these factors in any undertaking can have a detrimental effect on safety and safe operation.

In this book, we have discussed how safety regulations are not there just because they need to be followed—they are there to protect people from harm. This should be any organization's raison d'être; the underlying purpose of its undertaking. We have also examined how the "claim" of safety and the evidence that supports it—in terms of the documents, training, and so forth that we provide to support that claim—are only completely substantiated by appreciating and making allowances for the influences surrounding that evidence. These are the arguments around culture, operations, documentation, and administration, or CODA.

We submit that, by observing and understanding the factors that we have identified in CODA, that any organization will be much more able to not only comply with those regulations but also fundamentally ensure the safety of all those with whom the organization engages: staff; visitors; investors; neighbours; contractors; and so on. We hope that by viewing safety as a holistic function that affects, and is affected by, the *entirety* of an organization, will help to make you, and the place or places you work in, safer.

Works cited

AAIB, (1992). *Report on the accident to BAC One-Eleven, G-B JRT over Didcot, Oxfordshire on 10 June 1990.* Air Accidents Investigation Branch, Department of Transport. London: Crown copyright. ISBN 0-11-551099-0.

AAIB, (2003). *Report on the Accident to Boeing 747-2B5F, HL-7451 near London Stansted Airport on 22 December 1999.* Air Accidents Investigation Board, Department of Transport. London: Crown Copyright.

AAIB. (1990). *Air Accidents Investigation Branch: Report on the accident to Boeing 737-400 G-OBME near Kegworth, Leicester on 8 January 1989.* London: HMSO.

Advisory Board on the Investigation of Suspension Bridges. (1944). The Failure of the Tacoma Narrows Bridge: A Reprint of Original Reports. Texas A&M College Engineering Experiment Station, Bulletin No. 78.

Ambrose, J. (2021, January 28). *UK electricity from renewables outpaces gas and coal power.* Retrieved 14 October 2021, from The Guardian: https://www.theguardian.com/environment/2021/jan/28/uk-electricity-from-renewables-outpaces-gas-and-coal-power

Benzene. (2017). *IARC monographs on the evaluation of carcinogenic risks to humans. 120.* Lyon: International Agency for Research on Cancer. Retrieved from Benzene IARC Monographs on the Evaluation of Carcinogenic Risks to Humans Volume 120.

Bloom, B.S. (1956). *Taxonomy of educational objectives, handbook I: The cognitive domain.* New York: David McKay Co Inc.

BP. (2010, September 8). *Deepwater Horizon Accident Investigation Report.* Retrieved 20 November 2012, from BP: http://www.bp.com/liveassets/bp_internet/globalbp/globalbp_uk_english/incident_response/STAGING/local_assets/downloads_pdfs/Deepwater_Horizon_Accident_Investigation_Report.pdf

Brendebach, B. (2017, July 26). *Decommissioning of nuclear facilities: Germany's experience.* Retrieved 13 January 2022, from International Atomic Energy Agency IAEA: https://www.iaea.org/newscenter/news/decommissioning-of-nuclear-facilities-germanys-experience

Browning, J. B. (1993). *Union Carbide: Disaster at Bhopal.* Union Carbide Corporation. Retrieved 2 January 2013, from http://www.bhopal.com/~/media/Files/Bhopal/browning.pdf

BIM. (2012). *Building information modelling.* London: Crown Copyright.

Chrissis, M.B., Konrad, M., & Shrum, S. (2011). *CMMI for development: Guidelines for process integration and product improvement (SEI series in software engineering).* Boston, MA: Addison-Wesley.

Collins, J. (2001). *Good to great.* London: Arrow Books Ltd.

Companies Act 2006. (2006). Retrieved from Legislation.gov.uk: https://www.legislation.gov.uk/ukpga/2006/46/part/10/chapter/2

Construction (Design and Management) Regulations. (2015). Retrieved from legislation.gov.uk: https://www.legislation.gov.uk/uksi/2015/51/contents/made

Council Directive 95/63/EC. (1995, December 5). Retrieved from Legislation.gov.uk: https://www.legislation.gov.uk/eudr/1995/63/adopted

Cornell, M. (2007, July 30). *What's your feed reading speed?* Retrieved from Matthew Cornell: http://www.matthewcornell.org/blog/2007/7/30/whats-your-feed-reading-speed.html#1

Crowe Horwath (2011). *Risk appetite and tolerance—guidance paper.* London: Institute of Risk Management, IRM.

Cullen, T. (1990). *The public inquiry into the Piper Alpha disaster.* Richmond: HM Stationary Office.

Dalzell, T. (2009). *The Routledge dictionary of modern American slang and unconventional English.* Abingdon: Taylor & Francis.

Directive 2009/104/EC of the European Parliament and of the Council. (2009, September 16). Retrieved from Legislation.gov.uk: https://www.legislation.gov.uk/eudr/2009/104

Directive 89/391/EEC—OSH "Framework Directive." (2021, May 3). Retrieved from European Agency for Safety and Health at Work: https://osha.europa.eu/en/legislation/directives/the-osh-framework-directive/11

Edwards v National Coal Board, 1 All ER 743 ((CA) 1949).

England, D., & Painting, A. (2022). *An effective strategy for safe design in engineering and construction.* Chichester: Wiley Blackwell.

FEMA. (2007). *I-35W bridge collapse and response—USFA-TR-166.* Minneapolis: FEMA.

Fischhoff, B., Lichtenstein, S., Slovic, P., Derby, S. L., & Keeney, R. (1981). *Acceptable risk.* Cambridge: Cambridge University Press.

García-Serna, J., Martínez, J. L., & Cocero, M. J. (2007). Green HAZOP analysis: Incorporating green engineering into design, assessment and implementation of chemical processes. *Green Chemistry, 9*(2), 111–124. doi:10.1039/b518092a

Germany's Power Mix 2020—Data, Charts & Key Findings. (2021, July 15). Retrieved 22 November 2021, from Strom-Report: https://strom-report.de/germany-power-generation-2020/

Ghanem Al Hashmi, W. S., & Arnold, B. (2021). *Governance and leadership in health and safety: A guide for board members and executive management.* Abingdon-on-Thames: Routledge.

Greiner, L. E. (1997, December 1). Evolution and revolution as organizations grow: A company's past has clues for management that are critical to future success. *Sage Journals, 10*(4), 397–409. Retrieved from: https://journals.sagepub.com/doi/10.1111/j.1741-6248.1997.00397.x

Hackitt, J. (2018). *Building a Safer Future—Final Report.* London: Crown Copright. Retrieved from: https://assets.publishing.service.gov.uk/government/uploads/system/uploads/attachment_data/file/707785/Building_a_Safer_Future_-_web.pdf

Haddon-Cave, C. (2009). *The Nimrod review.* London: The Stationary Office. Retrieved 14 November 2010, from: https://www.gov.uk/government/publications/the-nimrod-review

Hale, A. R., & Glendon, I. A. (1987). Individual behaviour in the control of danger. *Industrial Safety Series, 2.* ISBN 0-444-42838-0 (Vol. 2).

Hale, A. R., & Hale, M. (1970). Accidents in perspective. *Occupational Psychology, 44,* 115–121.

Health and Safety at Work—Summary statistics for Great Britain 2020. (2020). Health and Safety Executive. Retrieved from: https://www.hse.gov.uk/statistics/overall/hssh1920.pdf

Health And Safety at Work etc. Act. (1974). (HSE) Retrieved October 2020, from legislation.gov.uk: http://www.legislation.gov.uk/ukpga/1974/37

Henderson, R. (2015, June 8). *What gets measured gets done. Or does it?* Retrieved from Forbes: https://www.forbes.com/sites/ellevate/2015/06/08/what-gets-measured-gets-done-or-does-it/?sh=235a106513c8

Hollnagel, E. (2012). *FRAM: The functional resonance analysis method: Modelling complex socio-technical systems.* Boca Raton, FL: CRC Press. ISBN 978-1-409-44551-7.

HSE. (n.d.). *ALARP "at a glance."* Retrieved from Health and Safety Executive: https://www.hse.gov.uk/managing/theory/alarpglance.htm

HSG65—Managing for Health and Safety (Third ed.). (1997). HSE. ISBN 978-0-717-66456-6.

HSG65—Managing for Health and Safety. (2013). (HSE, Producer) Retrieved 14 October 2021, from Health and Safety Executive: https://www.hse.gov.uk/pubns/books/hsg65.htm

HSG245—Investigating Accidents and Incidents. (2004). London: HSE.

ISO 14001:2015—Environmental Management Systems—Requirements with Guidance for Use. (2015). Geneva: International Organization for Standardization. https://www.bsigroup.com/en-GB/iso-14001-environmental-management

ISO 9001:2015—Quality Management Systems—Requirements. (2015). International Organization for Standardization. Retrieved November 2020, from: https://www.bsigroup.com/en-GB/iso-9001/

Kaplan, L. (2003). Inuit snow terms: How many and what does it mean? In: Building capacity in arctic societies: Dynamics and shifting perspectives. In F. Trudel (Ed.), *2nd IPSSAS Seminar. Iqaluit.* Nunavut, Canada.

Kirkham, F. (2013). *Lakanal Coroner action plan—Appendix 1.* Retrieved 14 November 2021, from Southwark Council: https://moderngov.southwark.gov.uk/documents/s51824/Appendix%201%20Lakanal%20coroners%20recommendations.pdf

Knight, K. (2009, July 3). *Report to the Secretary of State by the Chief Fire and Rescue Adviser on the emerging issues arising from the fatal fire at Lakanal House, Camberwell.* Retrieved from Communities and Local Government: https://www.highrisefirefighting.co.uk/case/lacanal2009/1307046.pdf

Kurzgesagt. (2013, July 11). Mechanisms of evolution. Retrieved 23 July 2021, from: https://www.youtube.com/watch?v=hOfRNOKihOU

Labib, A. W., & Champaneri, R. (2012, May/June). The Bhopal disaster—learning from failures and evaluating risk. *Maintenance & Asset Management, 27*(3), 41–47. Retrieved from Maintenance Online: http://www.maintenanceonline.co.uk/maintenanceonline/content_images/Pages%2041,%2042,%2043,%2044,%2045,%2046,%2047.pdf

MCIB. (2013). *Cruise Ship COSTA CONCORDIA Marine casualty on January 13, 2012.* Marine Casualties Investigative Body. Rome: Ministry of Infrastructures and Transports.

MHCLG. (2018). *Building a safer future—independent review of building.* Ministry of Housing, Communities and Local Government. London: Crown Copyright.

NASA. (2012, May 4). Failure Modes, Effects, and Criticality Analysis (FMECA). Retrieved from: http://engineer.jpl.nasa.gov/practices.html

National statistics—Fire and Rescue Incident Statistics: England, Year Ending June 2021. (2021, November 11). (Home Office) Retrieved from Gov.uk: https://www.gov.uk/government/statistics/fire-and-rescue-incident-statistics-england-year-ending-june-2021/fire-and-rescue-incident-statistics-england-year-ending-june-2021

National statistics—Reported Road Casualties Great Britain, Provisional Results: 2020. (2021, June 24). (Home Office) Retrieved from Gov.uk: https://www.gov.uk/government/statistics/reported-road-casualties-great-britain-provisional-results-2020/reported-road-casualties-great-britain-provisional-results-2020

NTSB/AAR-14/01. (2013). *Descent below visual glidepath and impact with Seawall Asiana Airlines Flight 214 Boeing 777-200ER, HL7742 San Francisco, California*. Washington: National Transportation Safety Board.

Oil & Gas UK. (2008). Piper Alpha: Lessons Learnt, 2008. Retrieved 4 January 2013, from: http://www.oilandgasuk.co.uk/cmsfiles/modules/publications/pdfs/HS048.pdf

Parkinson, C. N. (1955). Parkinson's Law. *The Economist*. Retrieved 10 January 10 2022, from: http://www.berglas.org/Articles/parkinsons_law.pdf

PAS 91:2013+A1:2017—Construction Prequalification Questionnaires. (2017). London: bsigroup.

PAS 99:2012—Specification of Common Management System Requirements as a Framework for Integration. (2012). London: bsigroup.

PDSA Cycle. (2020). The W. Edwards Deming Institute. Retrieved from The Demming Institute: https://deming.org/explore/pdsa/

Pratchett, T. (1987). *Mort*. London: Gollancz. ISBN 0-552-13106-7.

Public Interest Disclosure Act 1998. (1998). Retrieved from Legislation.gov.uk: https://www.legislation.gov.uk/ukpga/1998/23

Rasmussen, J. (1982). Human errors. A taxonomy for describing human malfunction in industrial installations. *Journal of Occupational Accidents*, 4(2–4), 311–333. Retrieved from https://www.sciencedirect.com/science/article/pii/0376634982900414

Ratcliffe, S. (2016). *Oxford Essential Quotations* (4th ed.). Oxford: Oxford University Press.

Ritchie, H. (2020, February 10). *What are the safest and cleanest sources of energy?* Retrieved from Our World in Data: https://ourworldindata.org/safest-sources-of-energy

Road Safety Annual Report 2015. (2015). Retrieved 12 November 2021, from International Transport Forum: http://www.internationaltransportforum.org/Pub/pdf/15IRTAD_Summary.pdf

Rose, D. (2012, December 9). *Untold story of the Concorde disaster*. Retrieved 20 October 2021, from Ask the Pilot: https://askthepilot.com/untold-concorde-story/

Rosencranz, A. (1988). Bhopal, transnational corporations, and hazardous technologies. *Ambio*, 17(5), 336–341.

Safe use of work equipment—Provision and Use of Work Equipment Regulations 1998. Approved Code of Practice and guidance. (2014). Retrieved 14 January 2022, from Health and Safety Executive: https://www.hse.gov.uk/pubns/books/l22.htm

Serrat, O. (2017, May 23). The five whys technique. *Knowledge Solutions*, 307–310. https://doi.org/10.1007/978-981-10-0983-9_32

Sharman, A. (2016). *From accidents to zero: A practical guide to improving your workplace safety culture*. London: Routledge.

Summers (John) & Sons, Ltd v Frost [1955] 1 All E.R. 870 (1955).

Taleb, N. N. (2012). *Antifragile: Things that gain from disorder*. New York: Penguin.

The Boeing 737 MAX: Lessons for Engineering Ethics. (2020, July 10). Retrieved 16 October 2021, from National Library of Medicine: https://www.ncbi.nlm.nih.gov/pmc/articles/PMC7351545/

The Control of Major Accident Hazards Regulations. (2015). Retrieved 3 November 2021, from Legislation.gov.uk: https://www.legislation.gov.uk/uksi/2015/483/contents/made

The Health and Safety (Display Screen Equipment) Regulations. (1992). Retrieved 2 October 2021, from Legislation.gov.uk: https://www.legislation.gov.uk/uksi/1992/2792/contents

The Lifting Operations and Lifting Equipment Regulations. (1998). Retrieved 21 November 2021, from Legislation.gov.uk: https://www.legislation.gov.uk/uksi/1998/2307/contents/made

The Management of Health and Safety at Work Regulations. (1999). Retrieved from Legislation.gov.uk: https://www.legislation.gov.uk/uksi/1999/3242/contents

The Provision and Use of Work Equipment Regulations. (1998). Retrieved from Legislation.gov.uk: https://www.legislation.gov.uk/uksi/1998/2306/contents/made

The Workplace (Health, Safety and Welfare) Regulations. (1992). Retrieved from legislation.gov.uk: https://www.legislation.gov.uk/uksi/1992/3004/contents/made

US Senate. (2021). *Aviation Safety Whistleblower Report*. United States Senate Committee on Commerce, Science, and Transportation. https://www.commerce.senate.gov/services/files/48E3E2DE-6DFC-4602-BADF-8926F551B670

Werner, J. (2014, April 17). *The design flaw that almost wiped out an NYC skyscraper*. Retrieved 13 January 2022, from Slate: http://www.slate.com/blogs/the_eye/2014/04/17/the_citicorp_tower_design_flaw_that_could_have_wiped_out_the_skyscraper.html?via=gdpr-consent

What Is PRINCE2. (2022). Retrieved from AXELOS: https://www.axelos.com/certifications/propath/prince2-project-management

Index

Printed in the United States
by Baker & Taylor Publisher Services